IMAGES
of America

THE TUNBRIDGE WORLD'S FAIR

On the cover: Please see page 12. (Courtesy of Tunbridge Historical Society.)

IMAGES
of America

THE TUNBRIDGE WORLD'S FAIR

Euclid Farnham

ARCADIA
PUBLISHING

ISBN 978-0-7385-5664-2

Published by Arcadia Publishing
Charleston, South Carolina

Printed in the United States of America

Library of Congress Catalog Card Number: 2007942412

For all general information contact Arcadia Publishing at:
Telephone 843-853-2070
Fax 843-853-0044
E-mail sales@arcadiapublishing.com
For customer service and orders:
Toll-Free 1-888-313-2665

Visit us on the Internet at www.arcadiapublishing.com

Contents

ACKNOWLEDGMENTS

The Tunbridge World's Fair thrives today because of community families, each generation contributing members to the shareholders, directors, superintendents, interpreters, fair staff, and volunteers. What emerges in this photo album is a larger picture of what all these people produced each year, but there is little record of their individual portraits. This history is, in essence, their family photo album, their legacy of the "gem nestled in a natural amphitheater." Only more recently do we have incidental snapshots of people at work at the fair; therefore, people-pictures within current contexts are included to portray and honor that contribution.

Most of the photographs came from the Tunbridge Historical Society archives, balanced with what the fair archives have in an informal collection. Thank you to Nancy Howe for her support of the project, scouring sources for photographs and clippings. Dick Rogers for making his albums on recent fairs available, that is, his critical record of different areas at fair time. Thank you also to the many casual photographers who have sent pictures to the fair over the years. Over time many are unknown and the pictures have become part of the archives and accepted for possible use. We have given credit and sources wherever possible. Photographers requesting credit are acknowledged on their photographs.

Thank you to Arcadia Publishing for approaching me and presenting this opportunity to share the fair's history, something I have wanted to do.

The modern, empty fairground pictures were taken a week before the fair on a clear day, when the white buildings gleamed and the green roofs contrasted with the forested hills of late summer. The crew was putting finishing touches on the area, and it was a moment of pause before all the action and the crowds arrived for another fair season.

This book could not have seen the light of day without the hard work of Rebecca Béguin who scanned the photographs, helped with the arrangement of the material, pushed, and prodded. One can only say a humble "thank you."

Thanks also to Priscilla, my wife, who spent hours editing the written material. She is my inspiration and the reason the book has this kind of quality.

INTRODUCTION

Vermonters always have been competitive with their farm animals, wanting a local event to determine who had the fastest horse or best-looking cow. Starting in 1847 and continuing for at least 10 years, an Orange County Agricultural Day moved around the county from Bradford to Chelsea; however, many residents yearned for a permanent home for a fair.

In 1861, a fair was held on the Elisha Lougee farm in North Tunbridge. Six years later, the Tunbridge Agricultural Society was formed and a more structured fair was held at the Lougee place. Local men established a crude racetrack and built a pole shed. These one-day events always featured a guest speaker. In 1867, former Lt. Gov. Burnham Martin (perhaps trying to impress the locals) referred to the event as the "Little World's Fair." The fair organizers thought this was a great idea, so on the 1868 fair handbills, in large type, were the words "World's Fair."

By 1875, it was obvious that the fair needed larger and more useful grounds. It was at this time that Aaron King and David F. Chapman, two Tunbridge entrepreneurs, offered their two 10-acre bottom land fields just to the west of Tunbridge Village.

The use of the land came with several strings attached. The landlords charged $50 annual land rent and took all the receipts from the grandstand as well as all concession rentals. In addition they would not allow a fair to take place until after the corn grown on the land had been harvested; consequently, the fair had to be postponed until late September or early October. Since the fair only got admission fees, it was always financially strapped. It would be this way until 1917 after both King and Chapman had died.

In retrospect it seems miraculous that the fair survived those early years. Most years the directors could cover the bills; however, in 1880, 1885, and 1889, they had to borrow to meet expenses.

Those involved with the fair in 1875 laid out the trademark half-mile racetrack on the grounds and worked on the first buildings: Floral Hall and the main ticket office. Within a few years they added the first grandstand as well as horse stalls and the cattle barns (some brought from the old Woodstock fairgrounds). Now that the fair had a permanent home the fair itself was extended to two days. In 1876 a third day was added. Also that year the organization increased the number of directors to 25 men from 13 towns around Tunbridge, in order to get more people involved in the fair.

For the next 20 years, the annual fairs went well until disaster struck in 1897. During a bicycle race on the racetrack, a young cyclist swerved and ran into a sulky. He was badly injured, soon sued the society, and won. The fair was ordered to pay $2,644.28 in damages and costs. This was a huge sum for an organization with only $600 in its treasury. A meeting of the members was soon called and 50 people pledged $40 each to pay off the debt. They also decided to incorporate with the State of Vermont as the Union Agricultural Society, the World's Fair. One hundred shares at $5 each were then sold to pay bills. The first fair after this in 1902 saw a profit of $197.16.

In 1907, cool weather and heavy rain drowned out the fair. When the directors met to set dates for the next year they "voted to hold the fair October 6, 7, and 8, 1908, if we have one at all." They did. For the first time since its inception, the 1918 fair was cancelled due to World War I and the world-wide influenza epidemic.

The Antique Cabin Museum was added in 1929 giving the fair an important new dimension. Ed Flint collected agricultural equipment and household goods from local farm families. Since then it has become one of the best museums at any fair in the East.

The 1938 fair started in the rain on September 20. The situation quickly worsened the next day when the great hurricane of 1938 roared through Vermont. Livestock had to be removed from the grounds, tents blew down, the electricity failed, and many people panicked. The directors tried to reschedule events later in the week, but few people attended—they stayed at home with their own hurricane-related problems.

It was not until 1999 and Hurricane Floyd that the present community had a taste of what a hurricane was like, moving cattle out of the barns in the flood plane and calling off the fair for a day. The grounds have had their share of flooding from the First Branch of the White River, which runs along the north and western borders.

Even though the shadow of war hung over the country in 1941 the fair was a success; however, it was to be the last fair until 1946. Many local farm men were away serving in the armed forces. For four long years everyone at home was intent on winning World War II.

By the 1950s, drunkenness and "girlie shows" were standard at county fairs. The Tunbridge Fair was no exception. "Honest women and upright folks" did not go to the fair at night and many forbade their daughters from going for their own safety. Nighttime was for carousers, mostly men. In 1958, a zealous local pastor shut down the "girl show." She wanted to end the fair, calling it a "den of gambling, drinking and sex." The fair, however, continued its long run and the girls were back the next year.

Sundays were added to the fair in 1961. In 1967, another woman pastor chose to become part of the fair by inaugurating a "fair service of worship" in the village church, an event that still continues. For many years, a demolition derby was the feature on Sunday afternoons. Recently Sunday has become Family Day with bargain admissions for young families.

The 100th anniversary fair in 1971 included many special events of celebration including a great parade led by Gov. Dean Davis on his Morgan horse.

As the fair entered its second century, it had become an end-of-the-year event that was rowdy and unmanageable. After the 1978 fair, the board of directors decided to change the direction of the event. Alcohol consumption was curtailed, general camping banned, and the "girlie shows" were no longer welcome. The fair still retains the lingering reputation of those days in spite of 30 years of change. It took several years of financial losses. Eventually a family and community atmosphere was reestablished, renewing the celebration of livestock and harvests much as it was when farming families first took a few days off from hard work to picnic on the grounds.

The directors also undertook renovation and building projects, and hired a summer maintenance crew. In addition to the world's fair, the grounds began to host horse shows, camper rallies, weddings, and the Vermont History Expo. It has become a small business with a budget process, strategic planning, and insurance policies. We look forward to the 150th in 2021.

One

The Fairgrounds

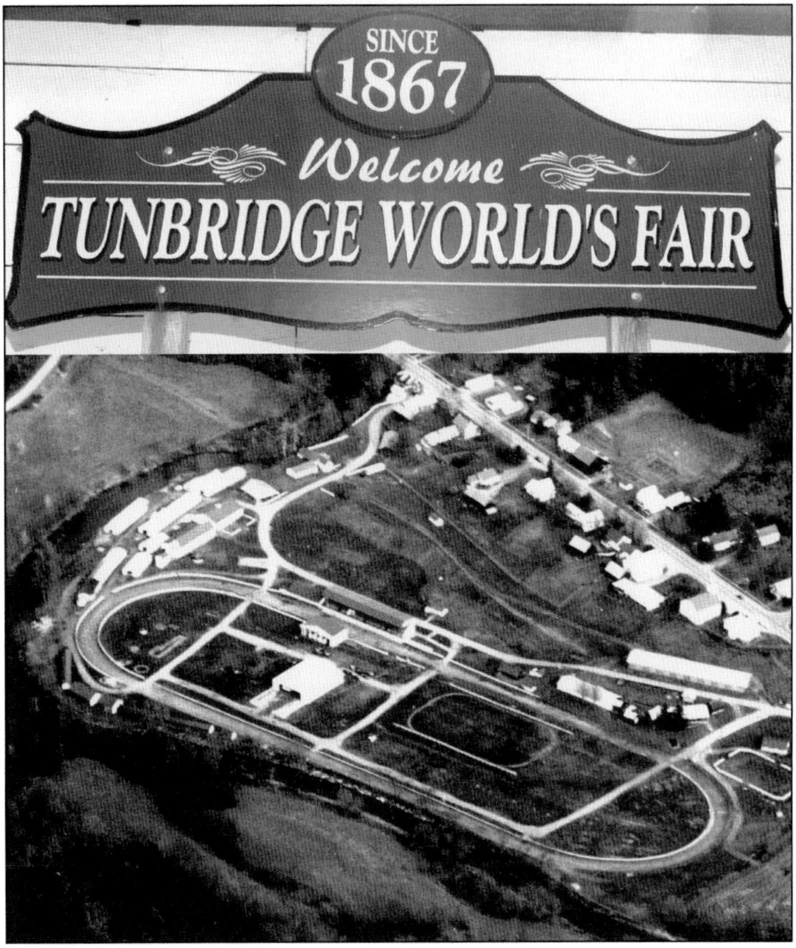

Visitors to the Tunbridge World's Fair are welcomed at the four main gates by signs that were first installed in the 1990s. An aerial view, right, to the east of Tunbridge Village and the fairgrounds was taken after the construction of the Dodge-Gilman Building. It clearly shows the proximity of the fair to the village; consequently, the fair directors have always sought to work closely with town officials. (Courtesy of Mark Dixon.)

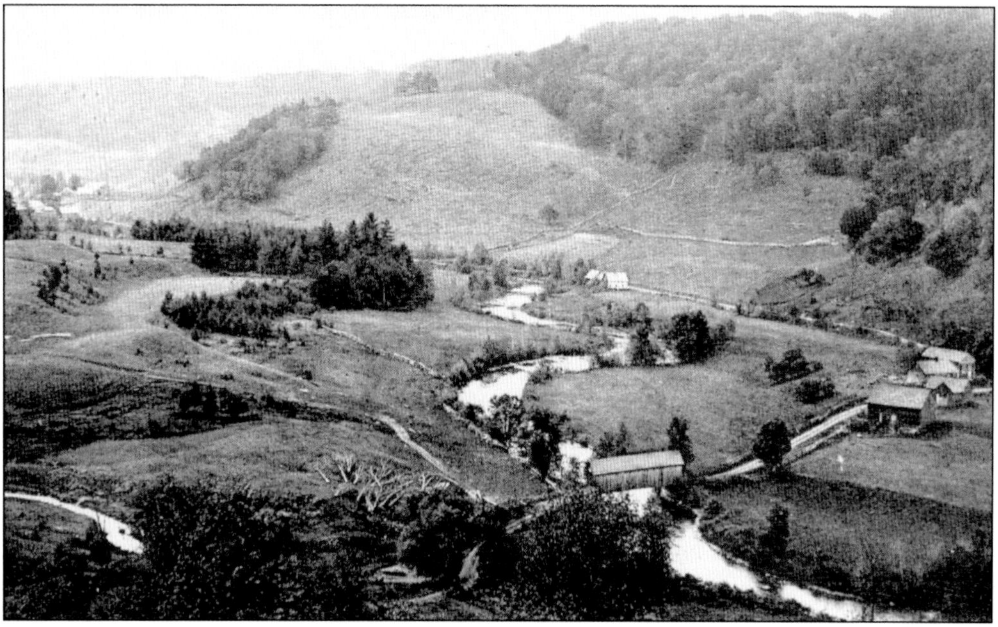

The fair started in 1867 in the fields on the left of the Larkin Covered Bridge in North Tunbridge. The property was owned by Elisha Lougee who allowed the fair organizers to lay out a crude track at the site. The event remained there until 1875. It was then that it was first called "the World's Fair."

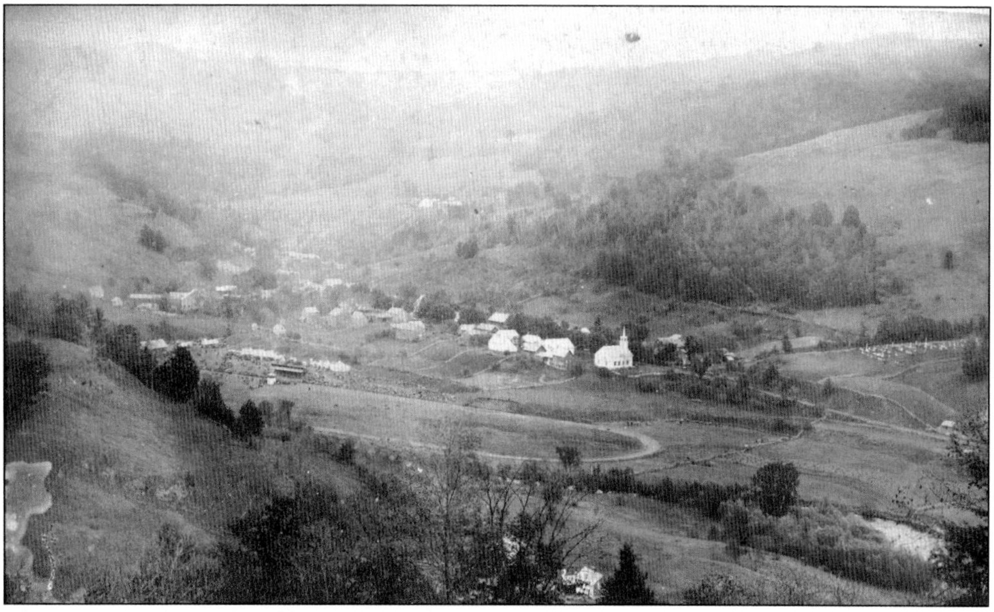

This scene, perhaps the oldest known photograph of the fair in its new location, dates from around 1880 (five years after the move). The first grandstand, the judge's stand, Floral Hall, and the Main Office were the first buildings. Each village homeowner had fences leading down the bank to the fair where they grew corn and pastured cattle. Below the church later became the site of Antique Hill. The tiny midway was just behind the grandstand.

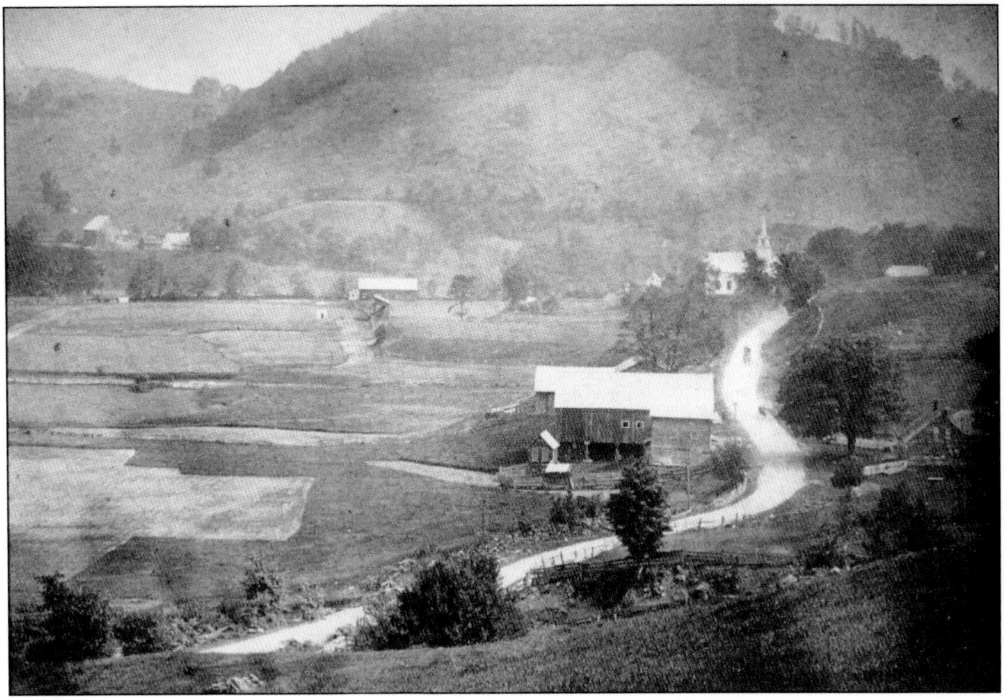

An 1890 misty view (from the south toward Sherlock's Farm) shows the dirt road (now Vermont Route 110) leading into Tunbridge Village. Decades of overgrazing by sheep cleared most of the hillsides. Only three buildings stand out: the grandstand, judge's stand, and Floral Hall. The racetrack cut through fields under cultivation.

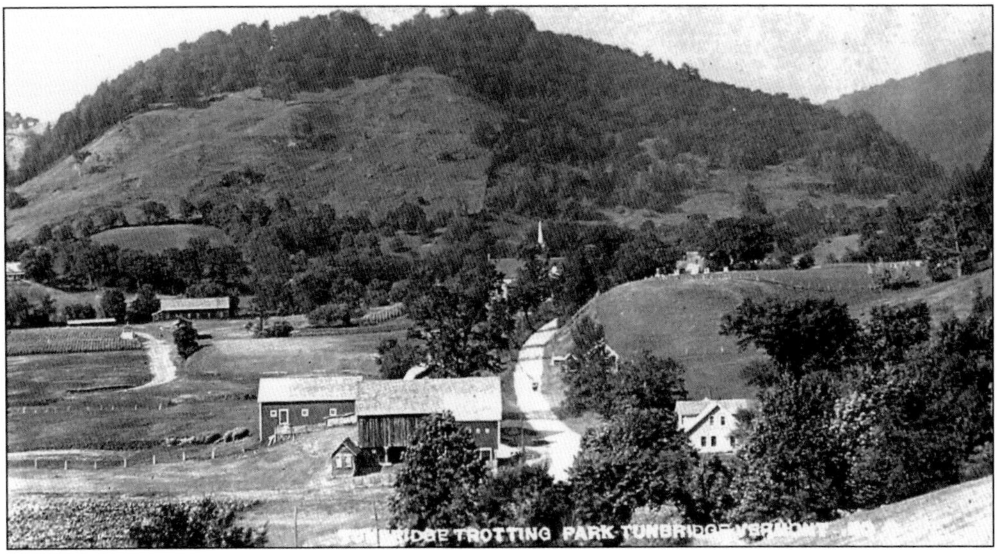

Ten years later several things stand out. The Tucker house had been painted and a dormer added. Above the house the cemetery is visible. On the fairgrounds corn was growing in the racetrack oval. The grounds were owned by the Aaron King and David Chapman families; consequently, the fair could not be held until October, after the landowners had harvested their corn.

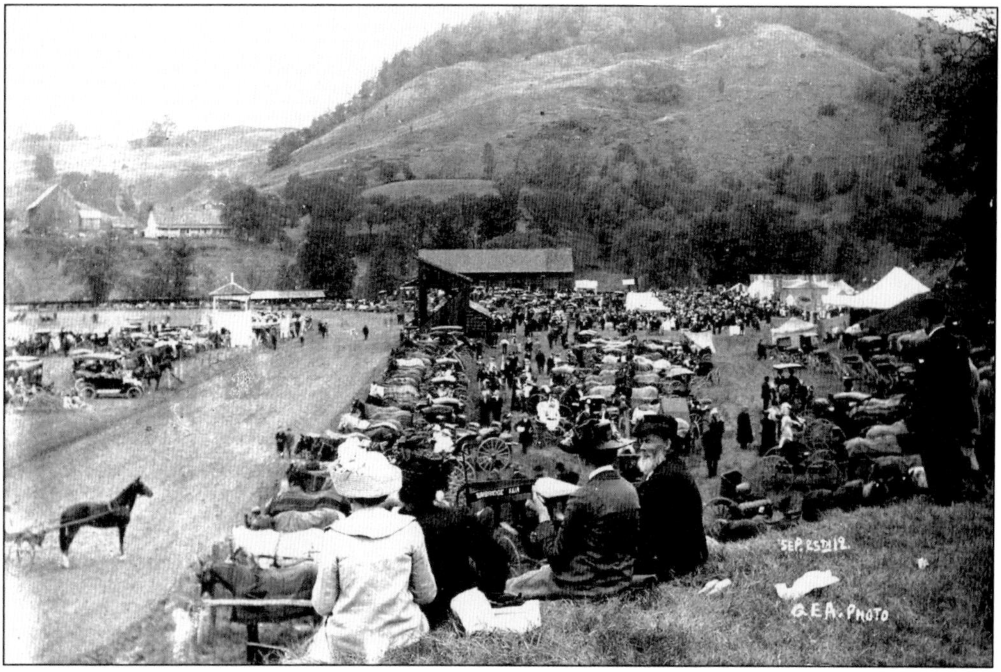

The people on the bank are holding a Tunbridge Fair banner in 1919. Looking north very few automobiles can be seen to the left of the track. The horse was still the main source of transportation. Beyond the resting horses is a crowd in the midway. Sheds are now visible at the north end of the track.

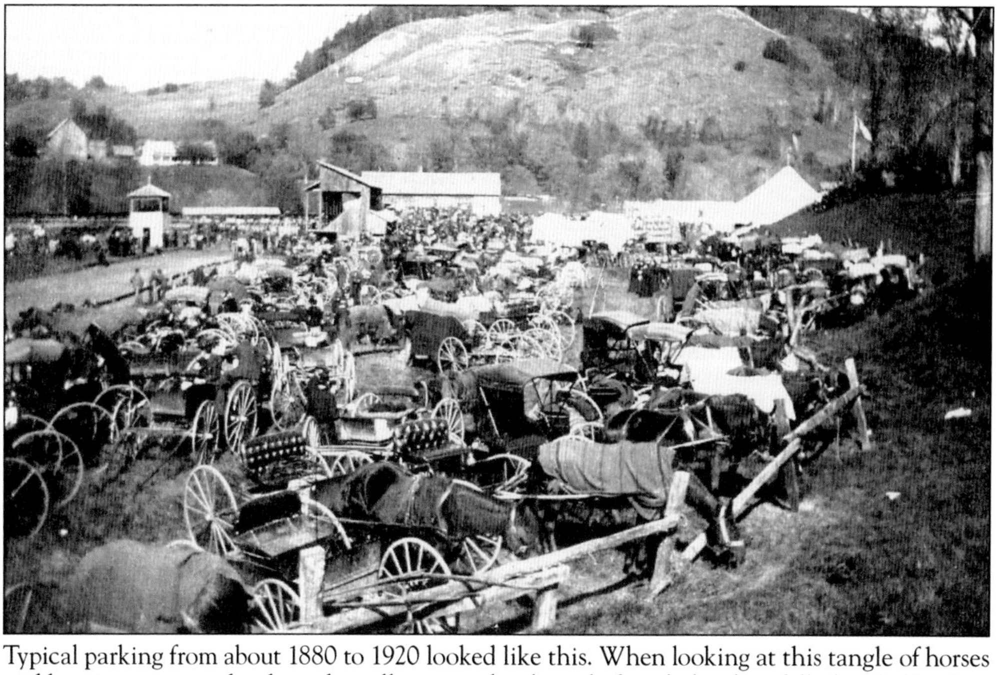

Typical parking from about 1880 to 1920 looked like this. When looking at this tangle of horses and buggies, one wonders how they all managed to leave before dark, when folks headed back to their farm chores. At the far right in the distance stands the carousel, which was the first ride to appear at the fair.

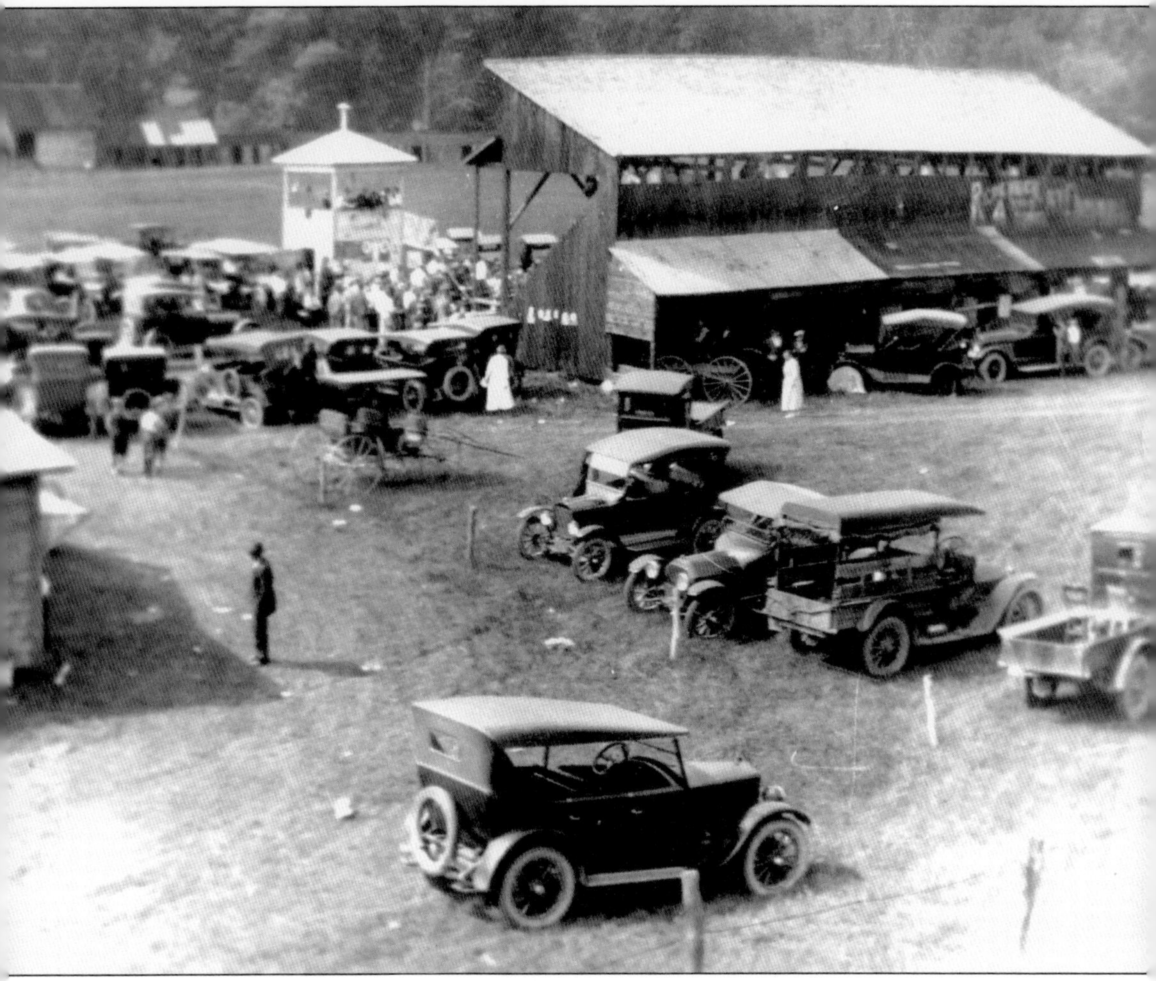

A closer view of the east side of the grandstand in the early 1920s reveals a business sign. Not only did it have a balcony, but several vendor booths attached to the back. This may have been early in the fair because it seems to be a slow day.

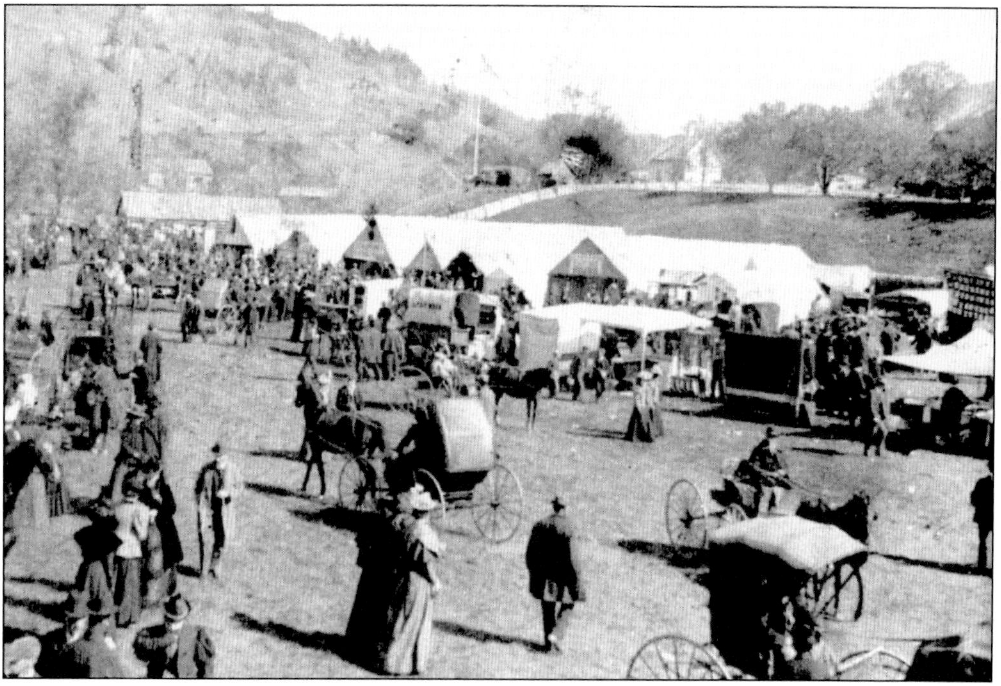

Apparently liability insurance was not a problem on the midway in 1885. Horse, buggies, and pedestrians intermingled happily. Did the well-dressed ladies watch where they stepped? Note a vendor's covered wagon and the flag on top of the hill next to the Main Office (beyond the fence).

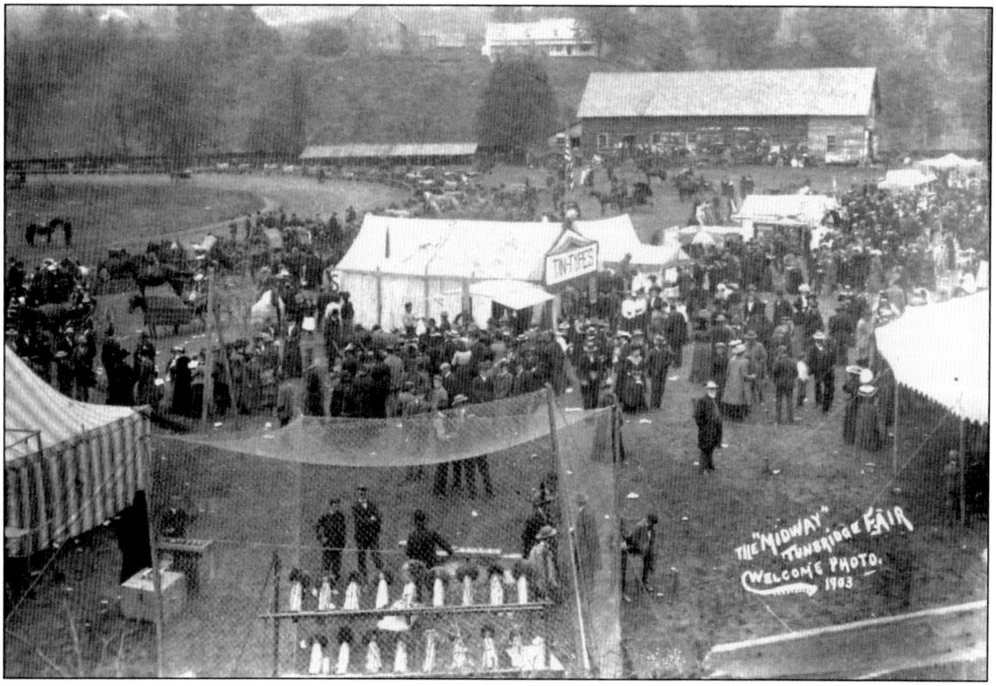

Mr. Welcome, a local photographer, took this picture of the midway in 1903. In the foreground is a baseball throwing game with its protective net. Tintypes were still very popular, as evidenced by the large tent and sign in the center. The midway is crowded with well-dressed fairgoers.

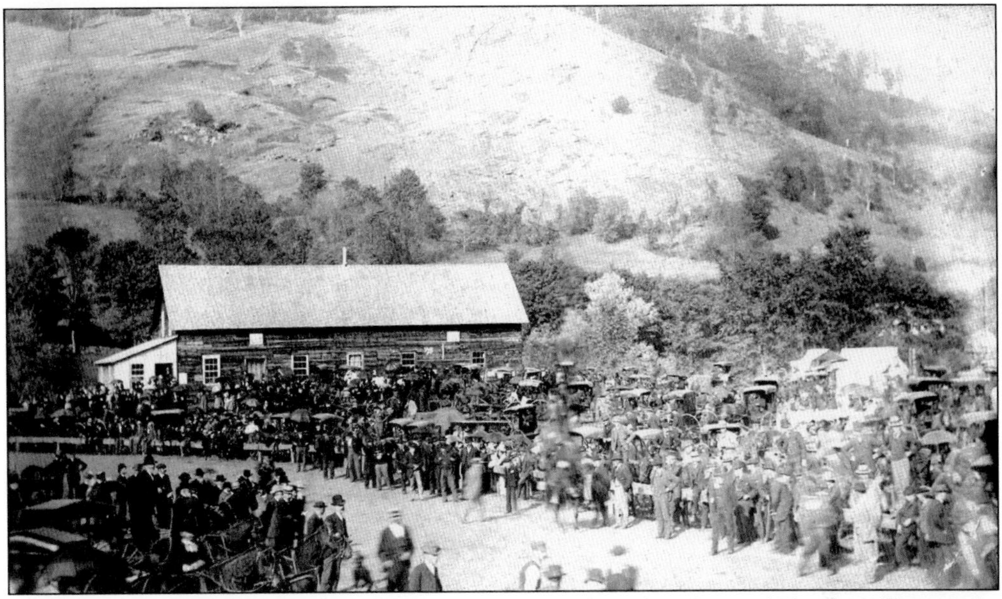

In 1890, the large crowd (at the north end of the track and in front of Floral Hall) waits for a horse race. There is no evidence of security at all. Soon after this, an accident did happen between a cyclist and a sulky. The lawsuit that followed almost put the fair out of business. A 1902 reorganization saved the event.

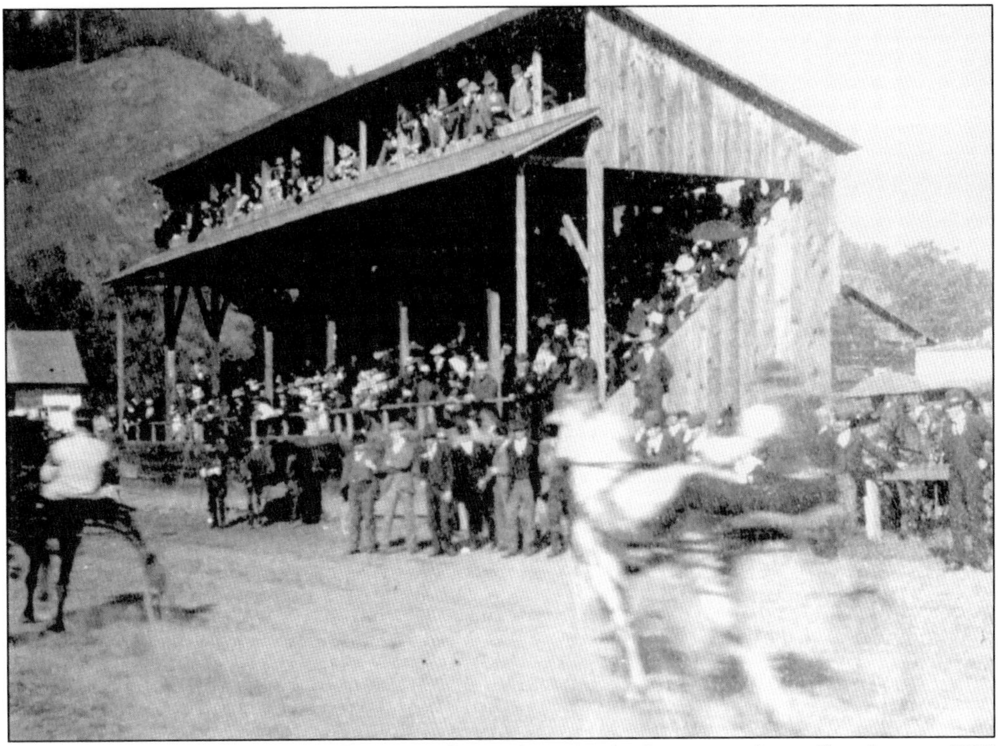

A 1900 harness race passes by the original grandstand, which was much smaller than its 1925 replacement. It did feature, however, a very high viewing area.

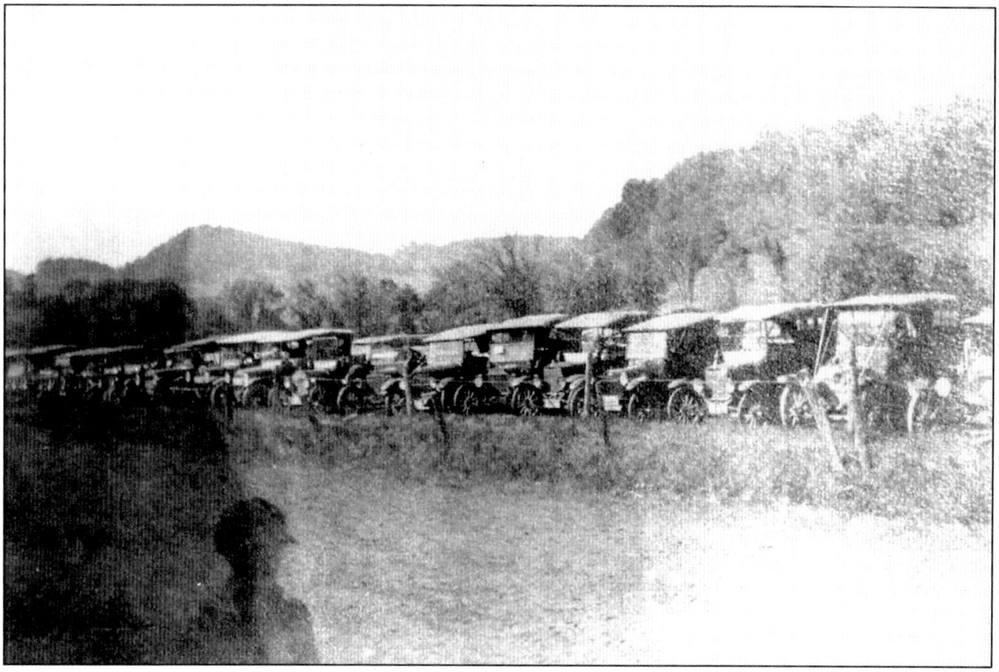

The automobile had arrived by 1920 (at the south bend of the track). They still looked a lot like horseless carriages. Behind a wire fence the cars were lined up neatly by the parking crew. The shadowy figure of a boy on the track shows his cap, which was typical of the times.

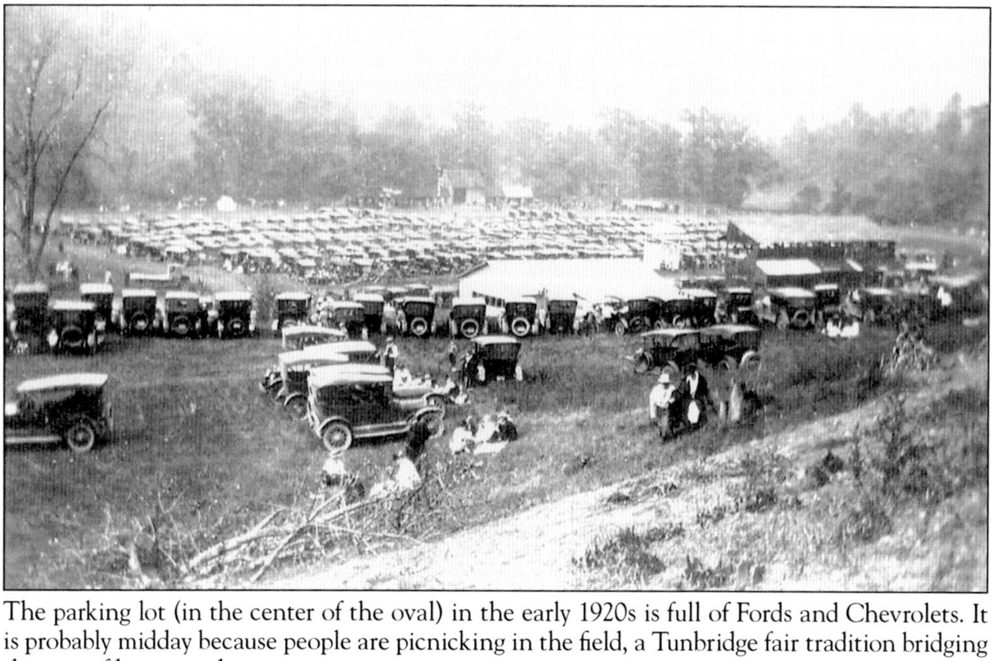

The parking lot (in the center of the oval) in the early 1920s is full of Fords and Chevrolets. It is probably midday because people are picnicking in the field, a Tunbridge fair tradition bridging the eras of horses and cars.

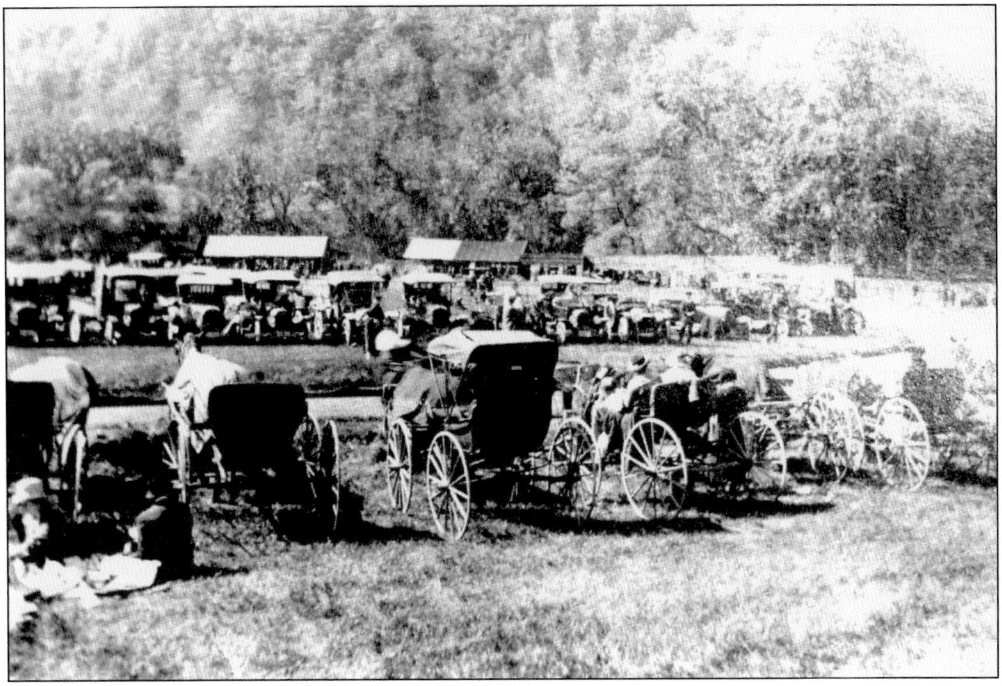

By the 1920s, parking on the north end showed a clear divide between automobiles and horse-drawn vehicles. On opposite sides of the track, they almost look like ancient army lines about to attack one another. A typical picnic is in progress on the left. In their open carriage, a family can be seen wearing their best hats to the fair.

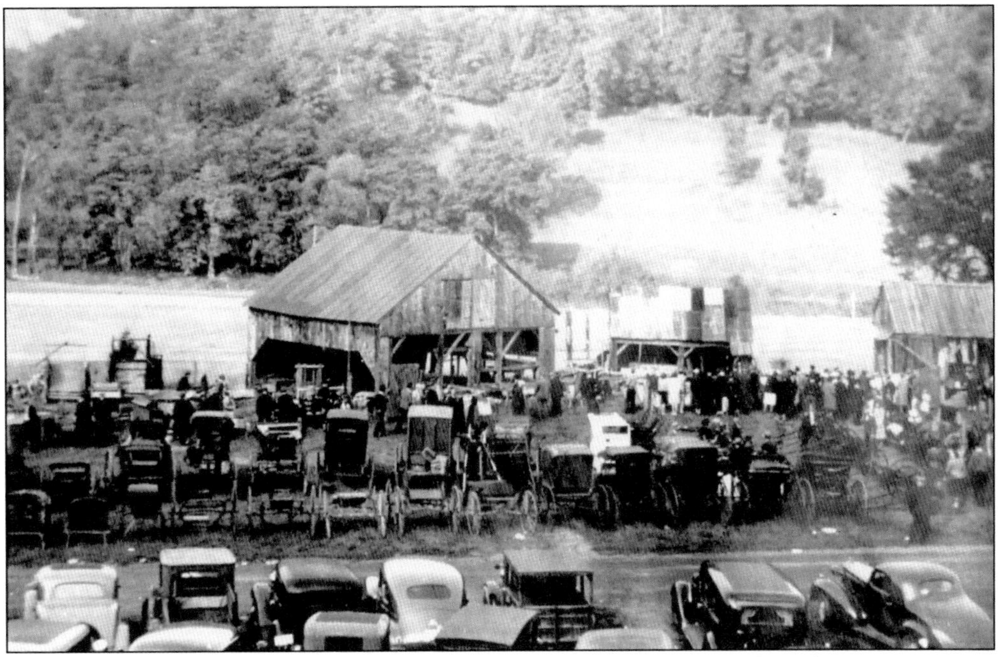

A row of carriages (now part of the fair collection) is arranged in front of new barns on Antique Hill (which opened in 1929). The track in the field beyond is empty because cars were parked across the street. To the left are horse-powered treadmills.

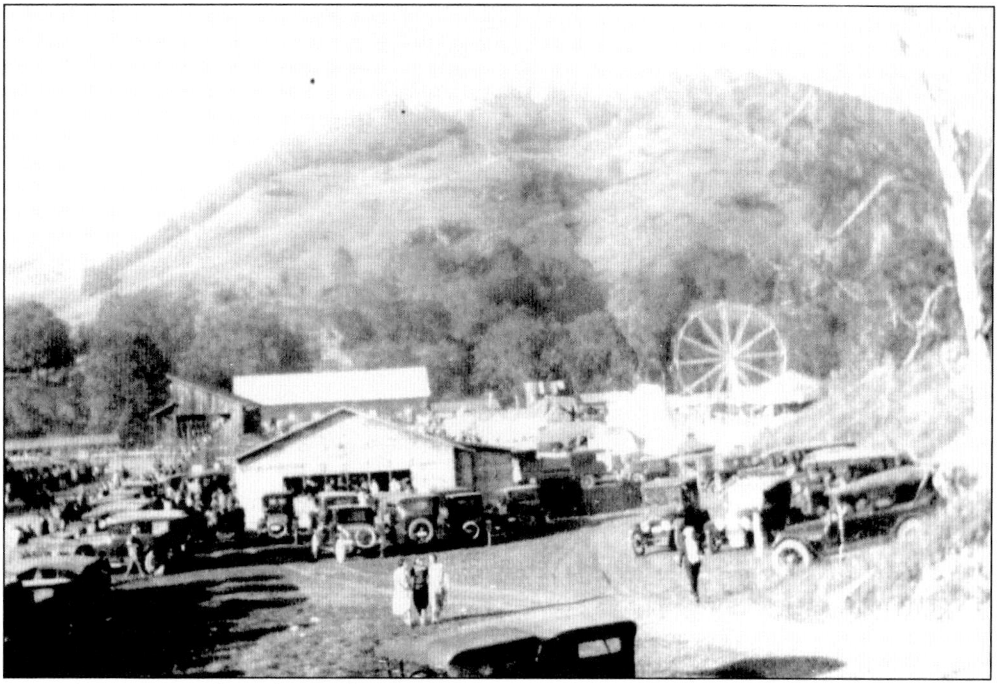

An early-1920s view from the south shows Dreamland, the new dance hall. Not long afterward, it was moved to the spot where the photographer stood to take this picture. The original grandstand still exists on the left, so this scene was prior to 1925. On the right are the merry-go-round and Ferris wheel.

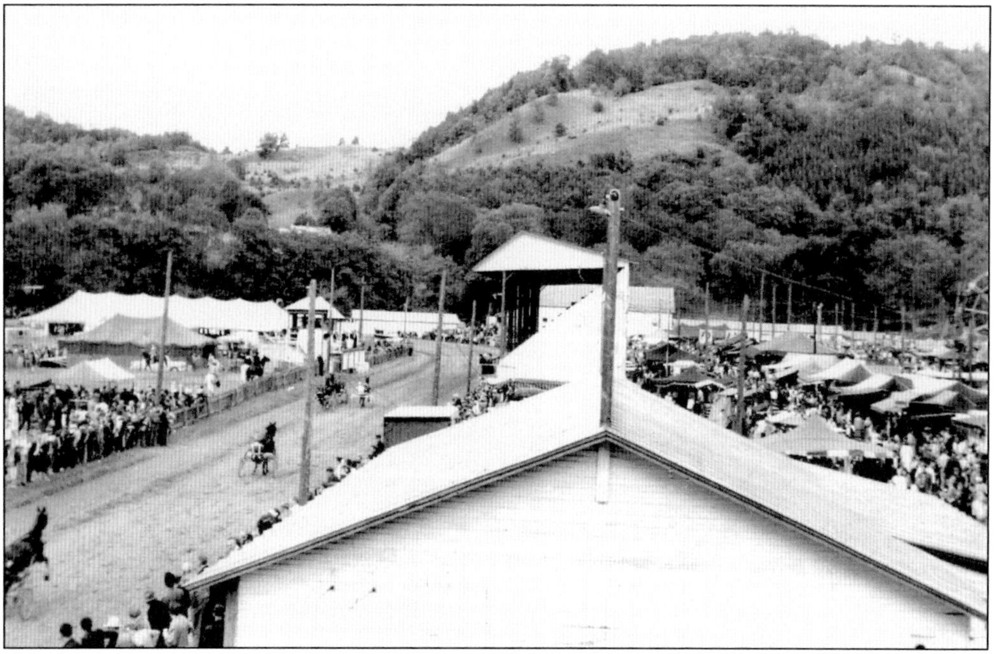

The roof of Dreamland is in the foreground. A race is heading down the home stretch. The fences along the racetrack, the midway, and Dreamland are crowded. This photograph was probably taken in the 1950s on a Wednesday (then the fair's most-popular day).

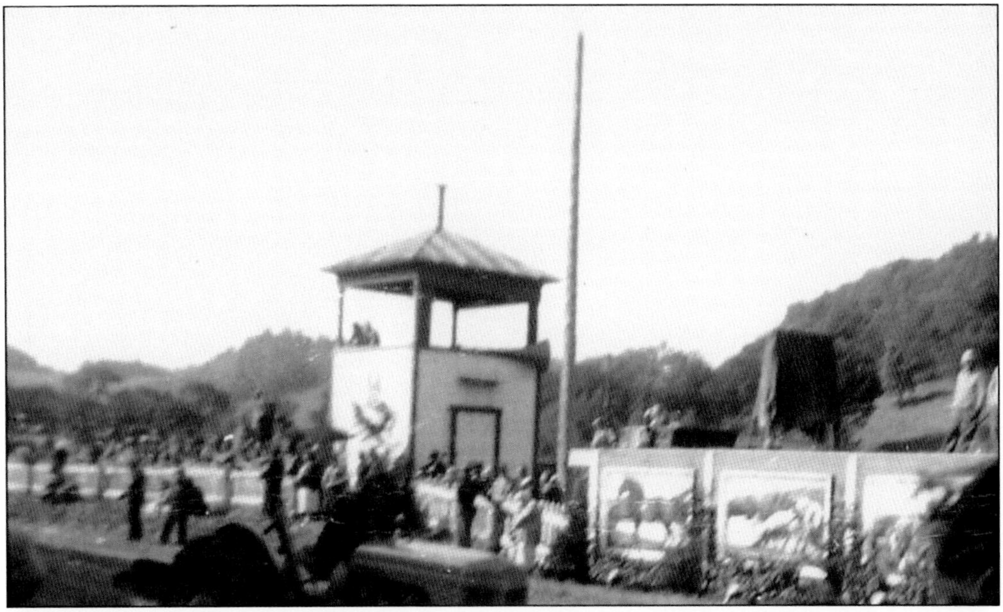

One of the first structures built on the new grounds was the judges' stand. Race officials needed an elevated position to monitor the half-mile racetrack, as well as the finish. Once it stood out, but it is now hidden by the stage roof. The paintings on the front of the open stage were done during the late 1940s. The new tractor in the foreground may be part of a parade.

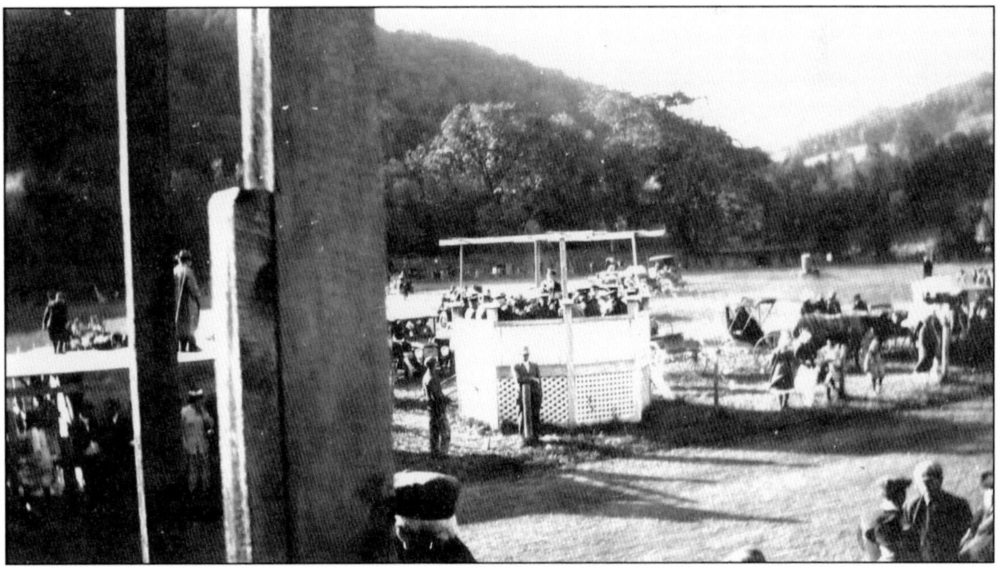

Music has always been popular during the fair, and many area town bands have been featured. In the 1920s, the newly built bandstand stood next to the stage for such performances. Eventually when the stage was covered, the bandstand was moved next to Floral Hall. It then became the gazebo.

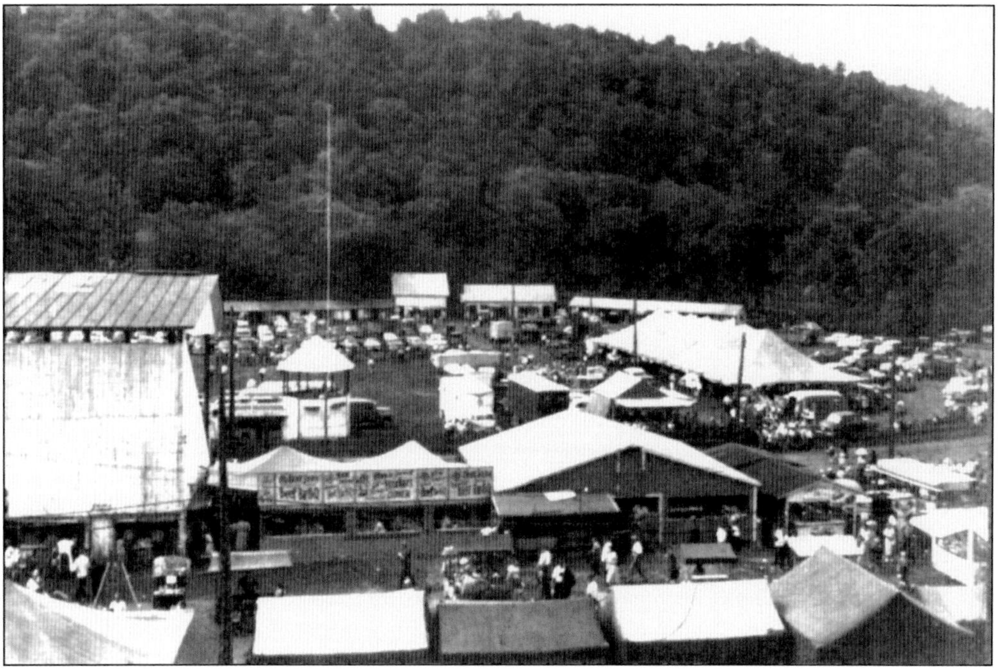

By the 1950s, the bandstand was covered. For years Wilmer Brooks owned a restaurant in Sharon. During fair week he opened Brooksie's on the fairgrounds. A canvas roof covered the left part and regular roofing was over the right section. The structure was changed piecemeal over the years. It is now used as the bingo hall and crew center.

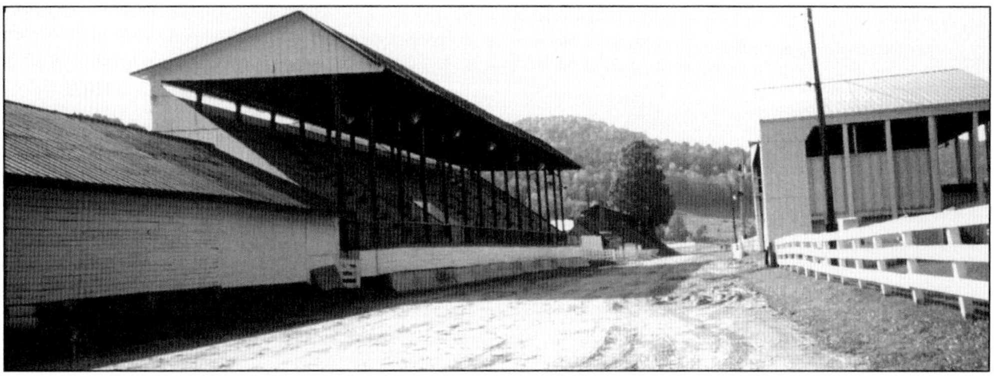

The home stretch viewed from the north shows the new stage across from the grandstand, which was built in 1925. For decades the stage was open. Various vaudeville acts, country artists, bands, and events were disrupted or cancelled by bad weather. In 1988, a local benefactor (unnamed) assisted the society in building a new stage.

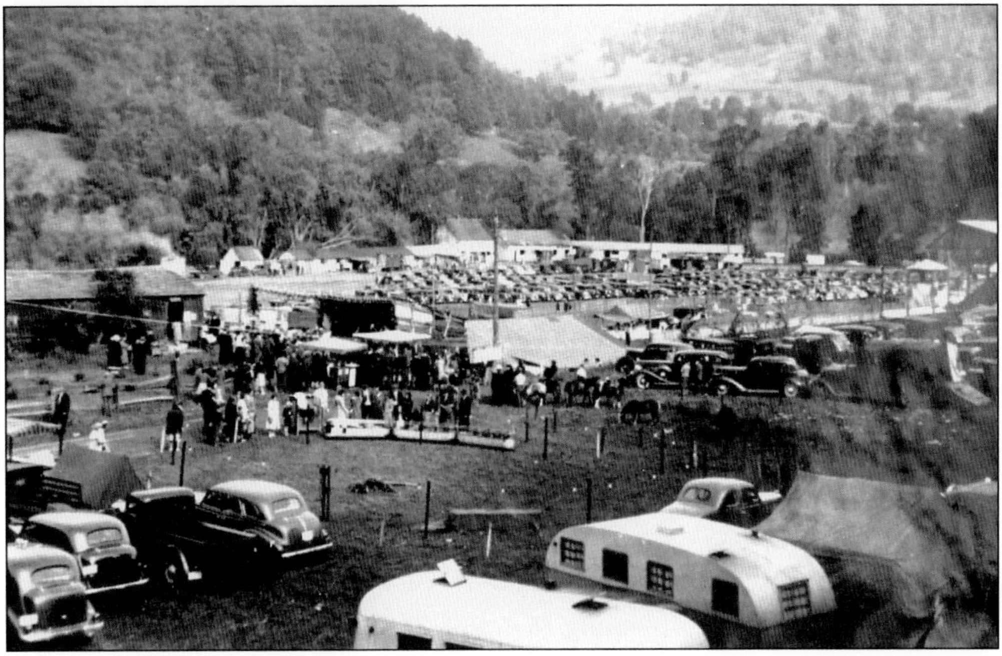

During the 1930s and 1940s, the little train in the center of this photograph was very popular with small children. For the first time in the 1940s, small campers were brought in by vendors. Antique Hill is to the left, ponies to ride are on the right, and the roof of Dreamland can be seen beyond them.

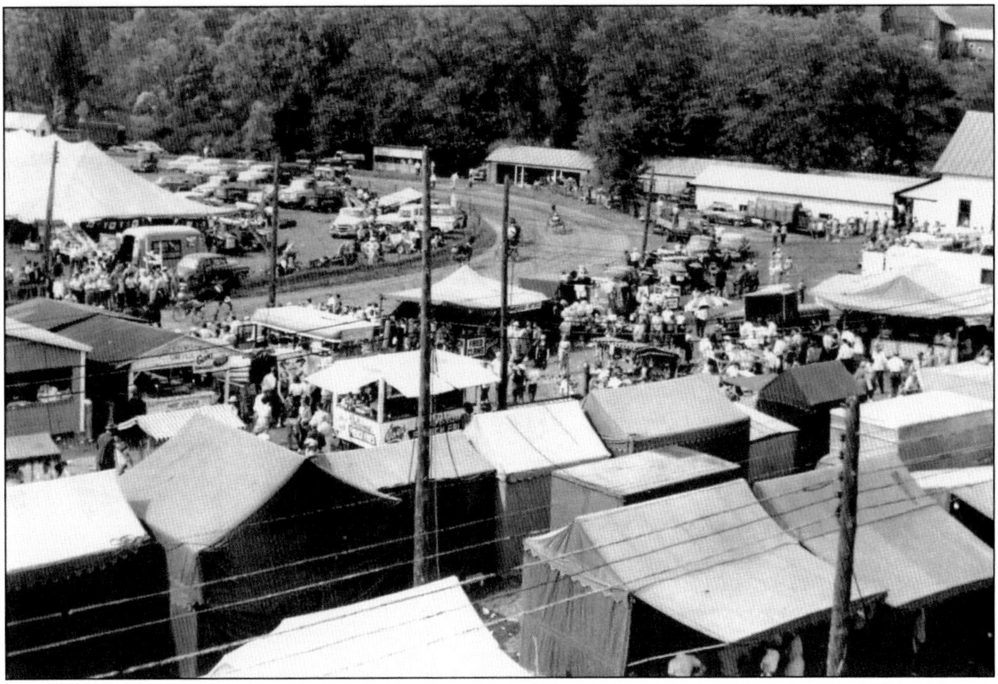

At the north end almost all concession stands still had canvas tents during the 1950s. Heavy rain and wind storms (such as the 1938 hurricane) had caused havoc and destroyed many of them. There is a race in progress. A big concessions tent is in the center of the oval.

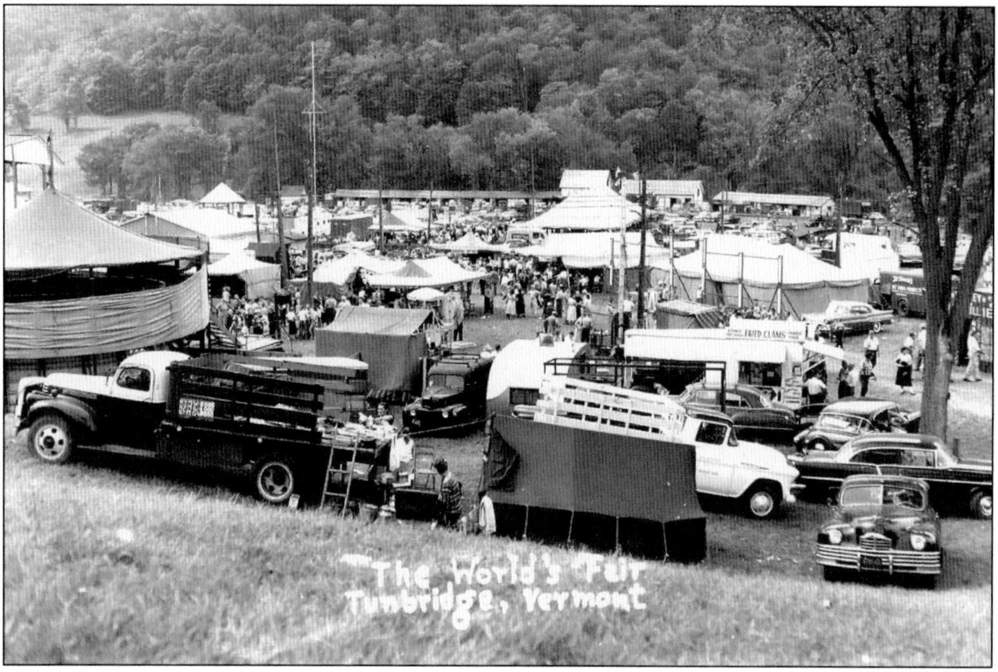

Another view of the north end in the 1950s shows an expanded and busy midway. The circular tent on the left features a motorcycle act with the driver going around the wooden sides of an inner ring at high speed. An exhibitor camped in a tent next to his truck.

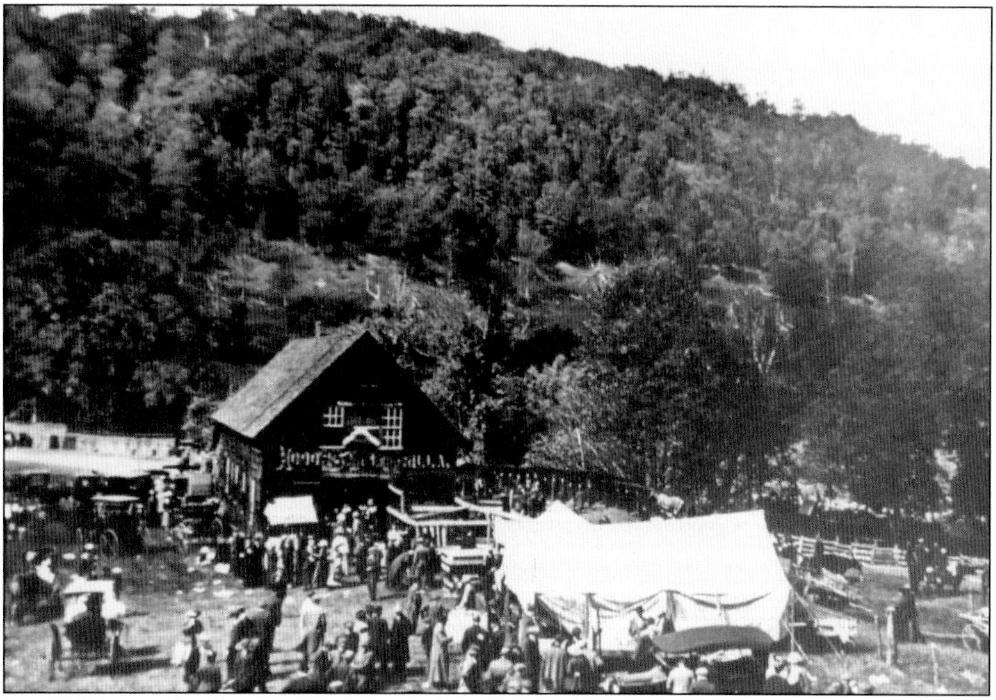

Floral Hall was the first new building on the fairgrounds after the 1875 move. While cattle and horse shows could be held outside, the various crafts and baking made by the ladies of the area could not. The hall was later enlarged to the size of the present-day building.

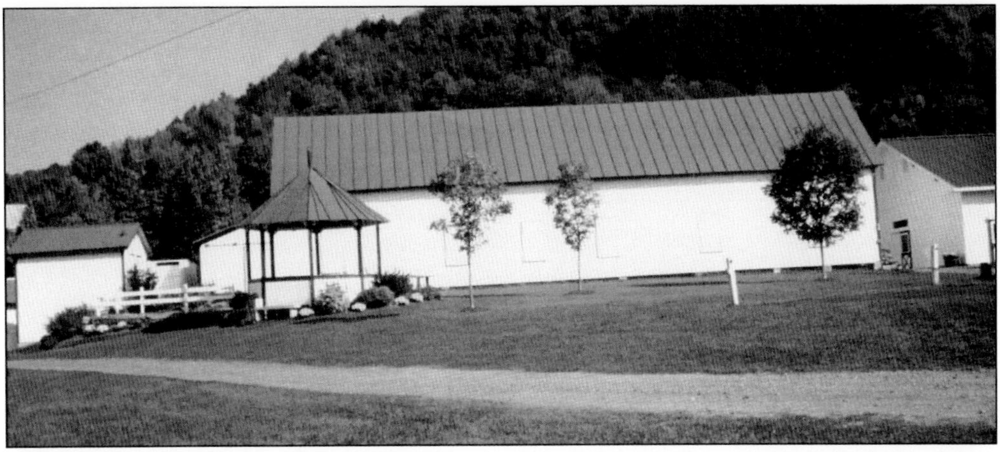

The bandstand was originally next to the racetrack and stage. For much of the fair's history, band concerts were popular. With the construction of the new stage in 1988, the bandstand was moved to this location next to Floral Hall. Now called the gazebo, it hosts radio broadcasts and informal music during the fair. Weddings are held there in the summer.

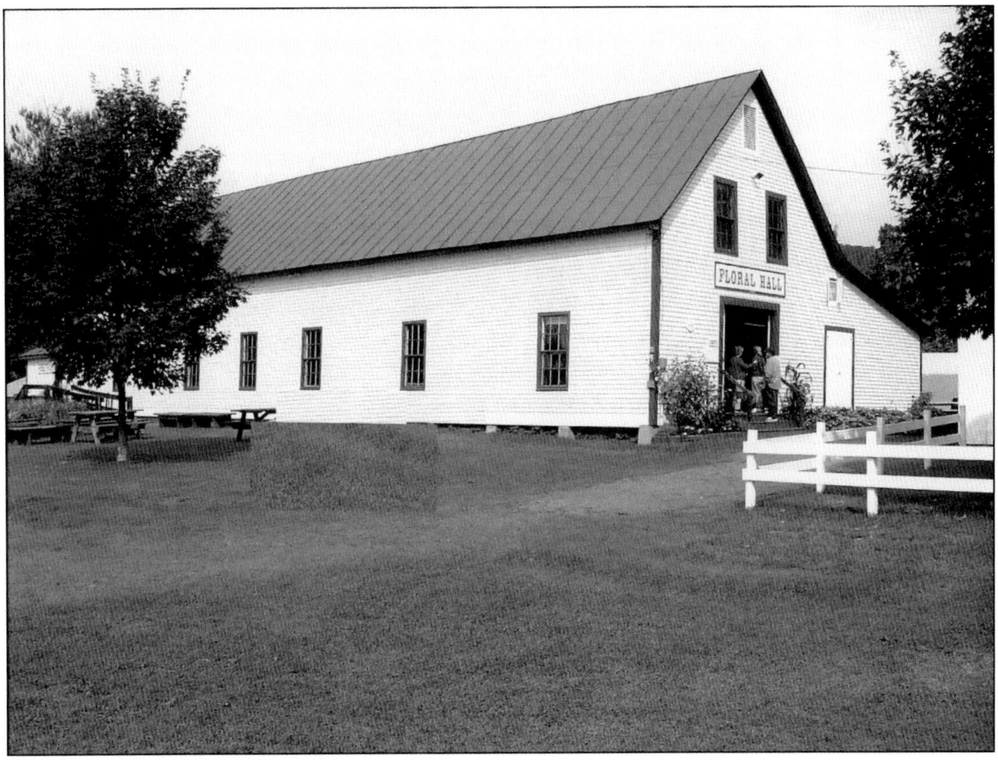

One can easily see that Floral Hall, over its 130-year existence, was built in three phases. The first section was the back (left). When the front section was added, windows were placed farther apart. Still later the ell section was built on the right. All fairgrounds buildings were painted uniformly white in the 1950s.

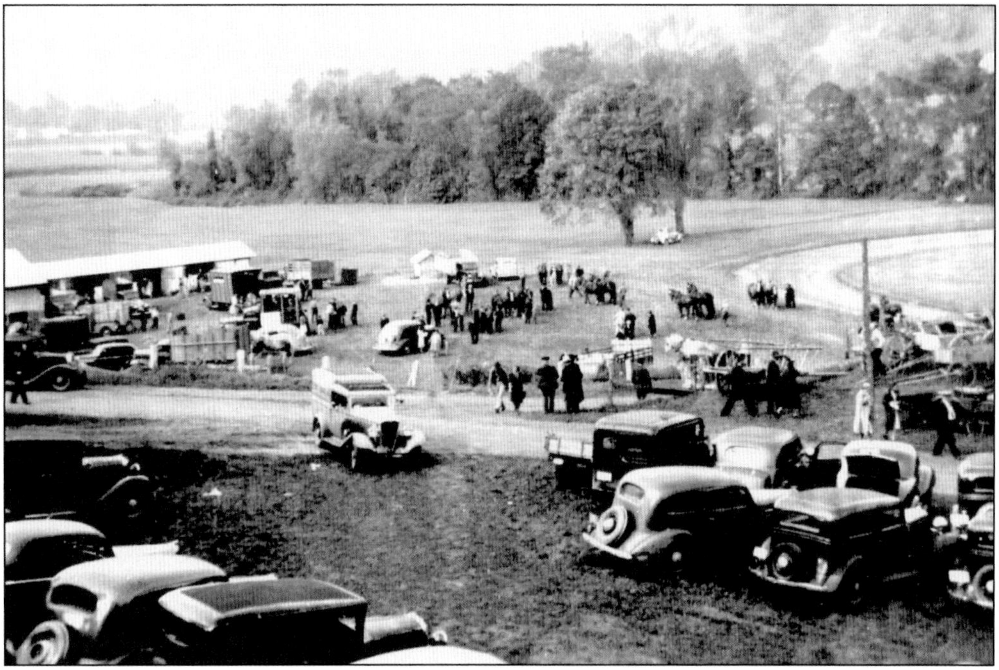

From Antique Hill looking south toward Tucker's Field is the horse show area after World War II (with barns forming the border). Two majestic trees stand where the main entrance is now. Teams of horses are to the right. The set up for the fair shows a mixture of trucks unloading and spectators milling around.

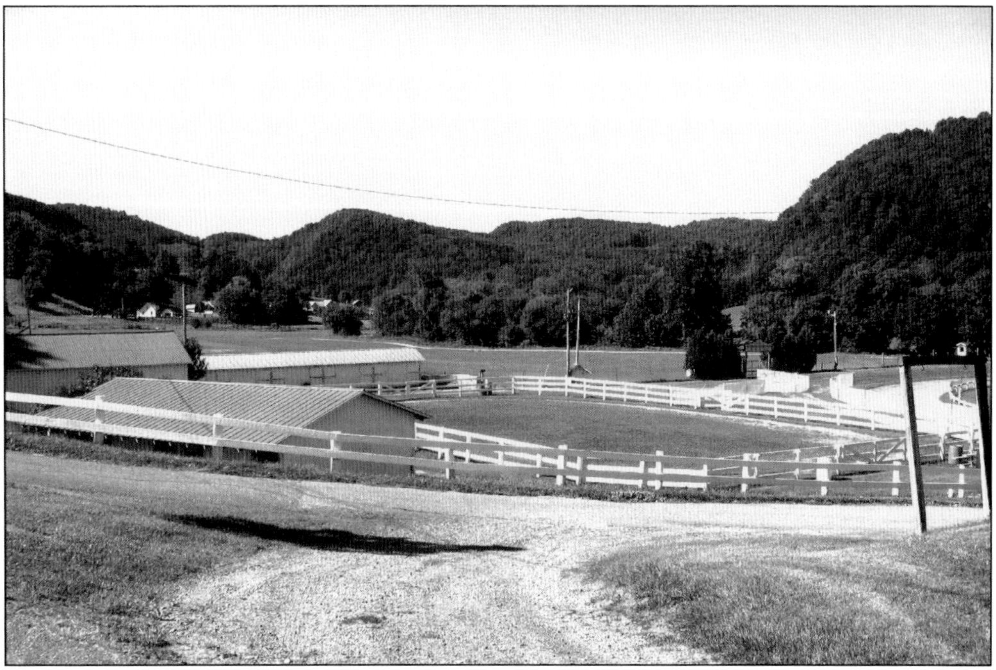

A view of the horse barns and arena shows how they appear today. Flooding over the banks of Vermont Route 110 during rain storms has caused silt build-up and muddy situations for horse exhibitors.

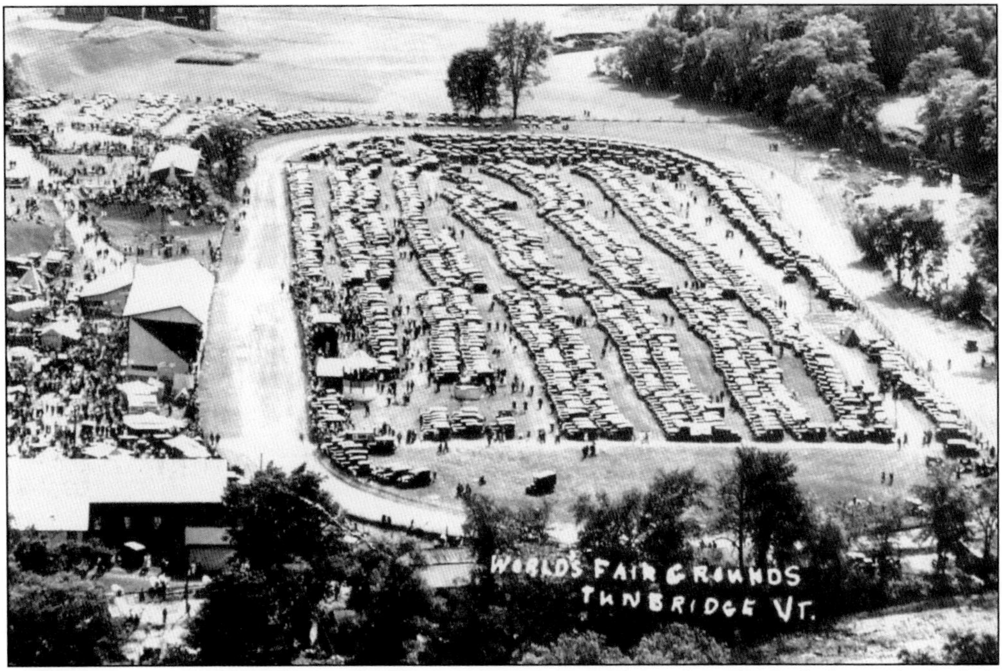

Parking was still in the center oval of the racetrack in 1946 (view from the hill above Sherlock's). Everyone came to the fair by car and there was no sign of a horse or buggy.

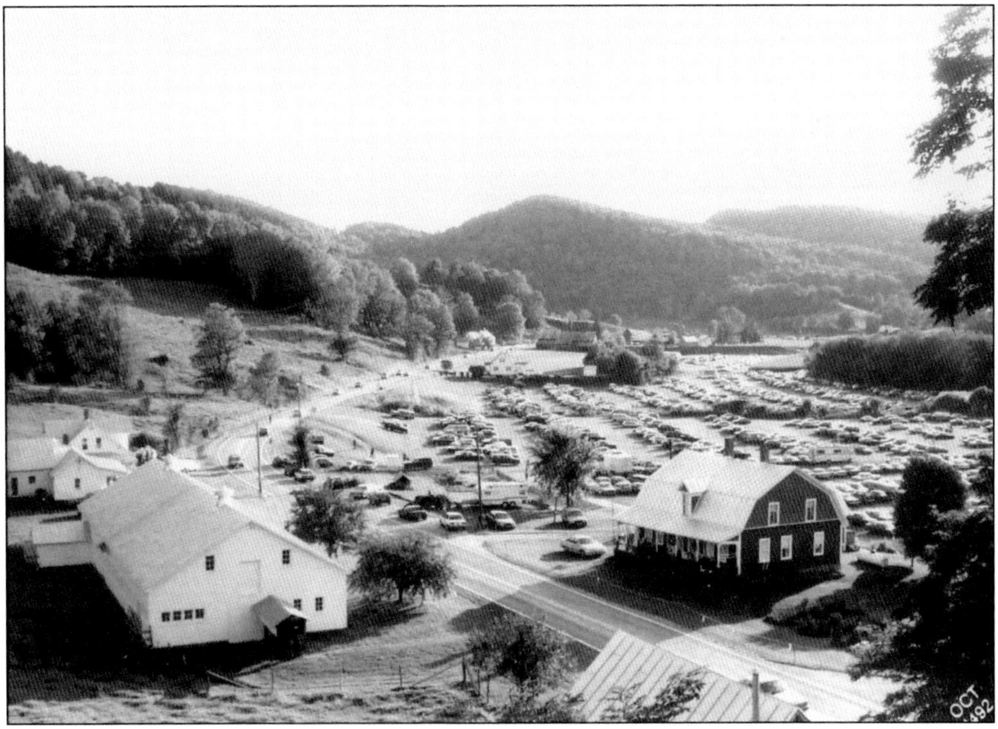

The fair began renting Tucker's Field for overflow parking. By 1992, cars were being parked as far as the Cilley Bridge. Fairgoers now may ride a shuttle to the south entrance.

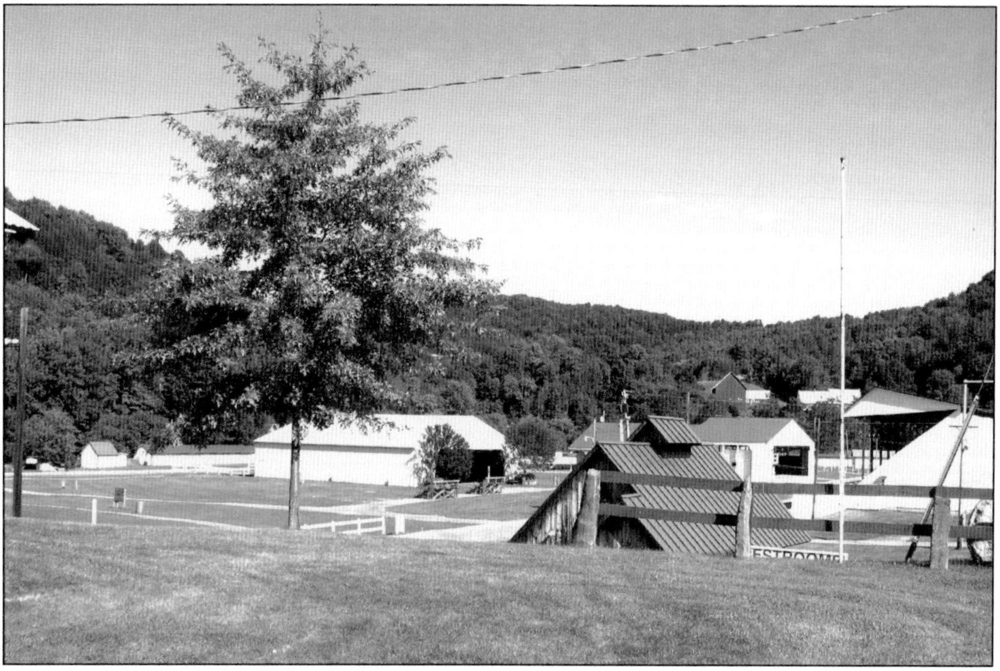

Several of the newer buildings can be seen from Antique Hill. From left to right are the pulling arena, a corner of the vendors' pavilion, the roof of the sugarhouse, and the covered stage.

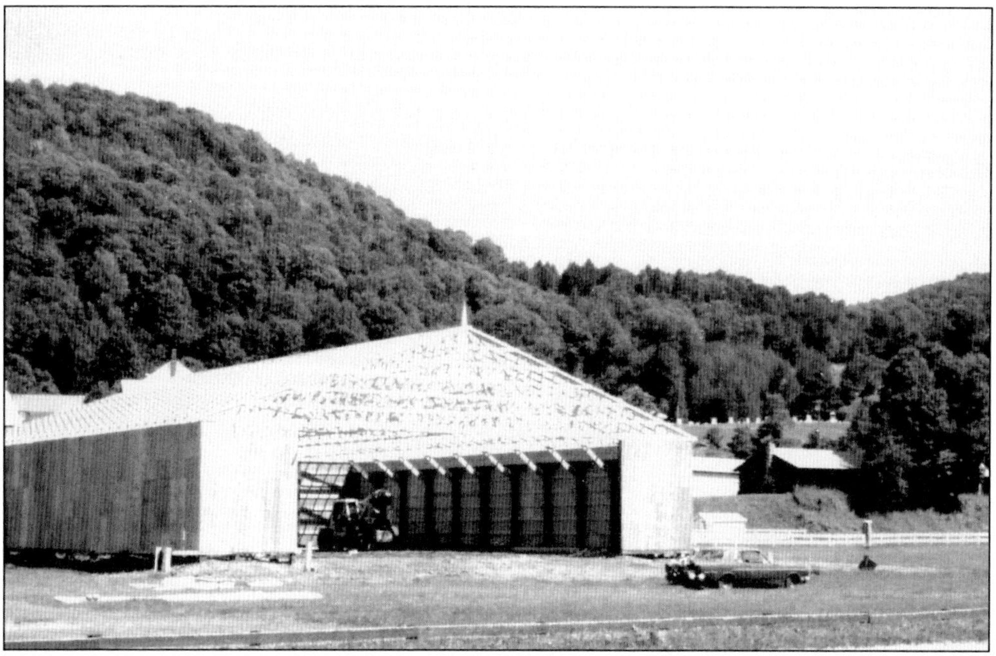

Since the beginning of the fair, horse, oxen, and pony pulling have been extremely popular. Pulling was done in the open for most of that time. Finally in 1989, the fair built a covered pulling arena with help from many donors; consequently, never will these events be cancelled because of inclement weather.

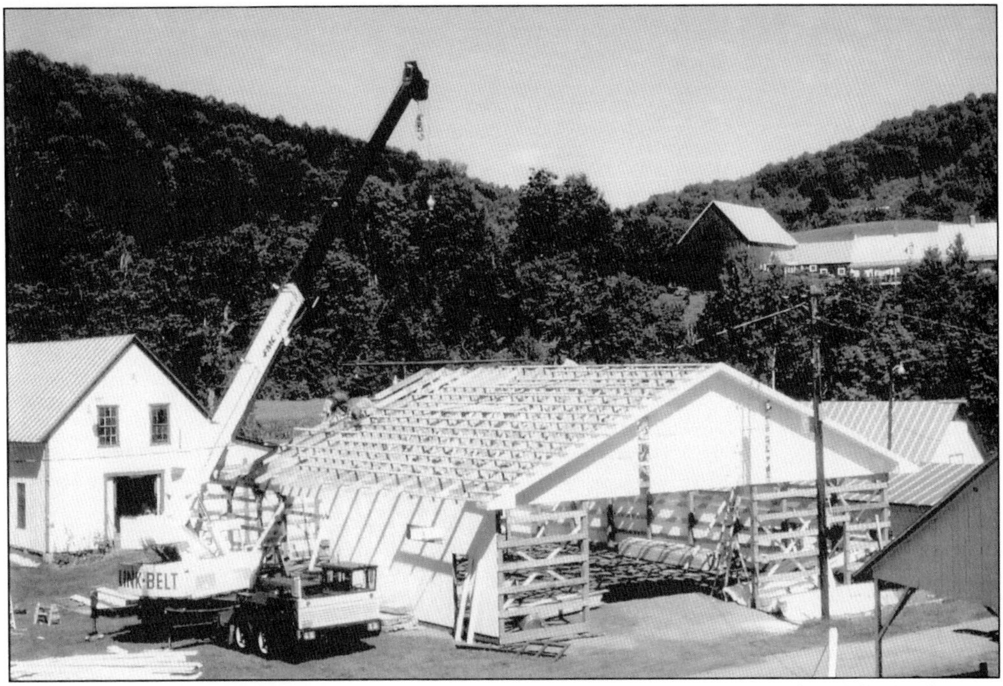

The Dodge-Gilman Building was constructed in 1995 to replace a large, outdated shed. The new building is now used during the whole season for weddings, reunions, meetings, and dinners. At the time of the fair, fruits and vegetables are displayed there.

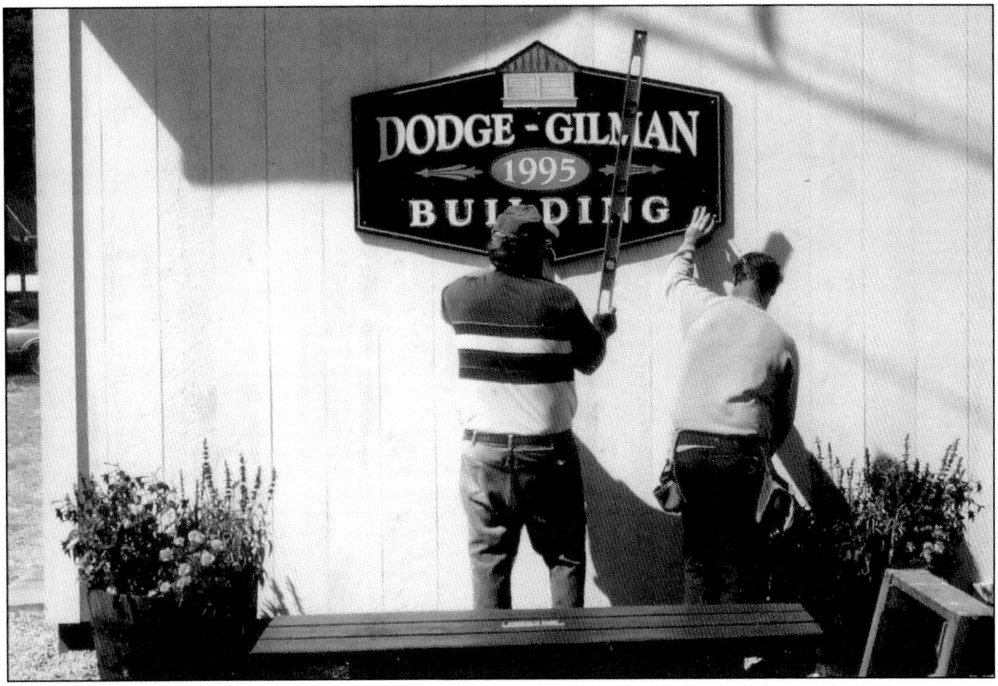

Dana Goodnow and Randy Tuller placed the sign on the side of the building. The Edgar Dodge and W. S. Gilman families contributed heavily toward the building's construction. A dedication and thank-you ceremony was held during the 1995 fair.

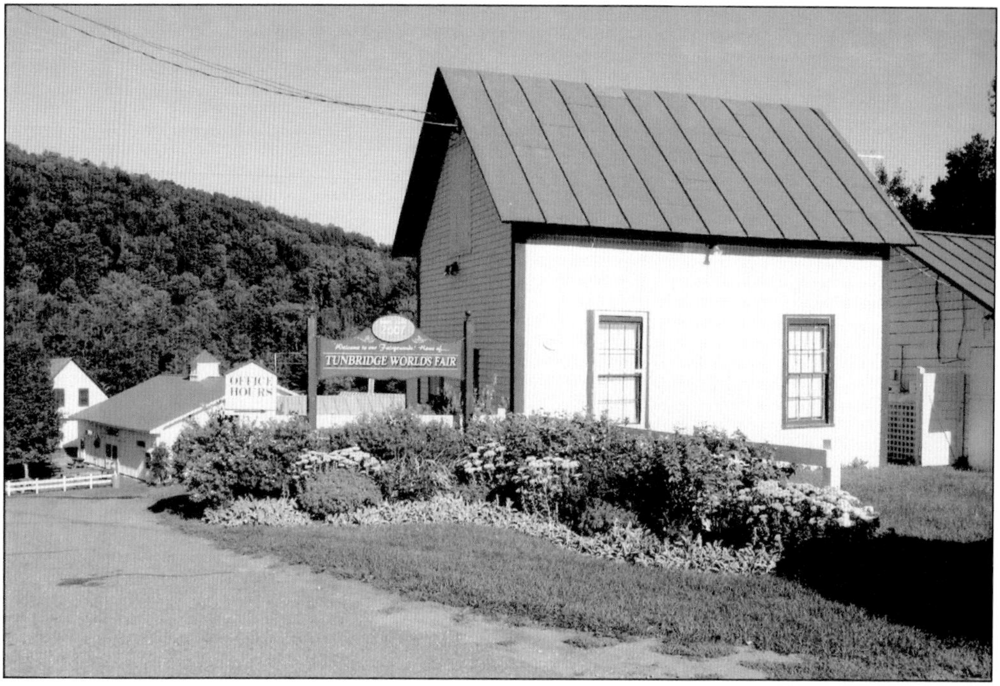

About 1880, this small farmhouse was moved to the fair and used as the main ticket office. It is now the president's year-round office. During the 1990s, a beautification program added new signs and gardens.

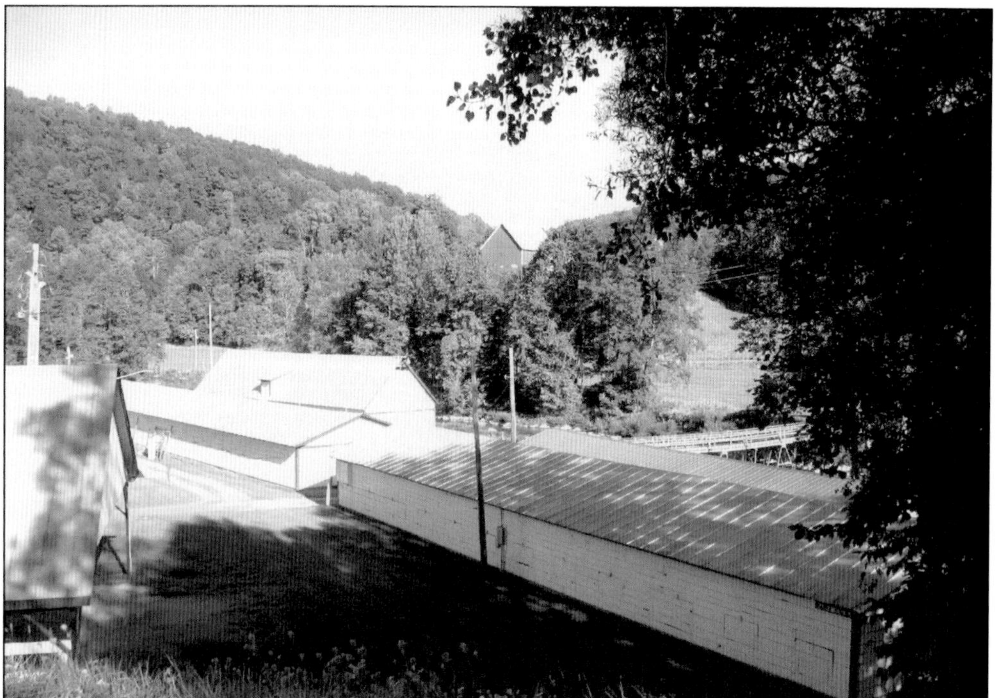

The edge of the judging arena is on the far left. The long cattle barns have signs designating the farm animals therein.

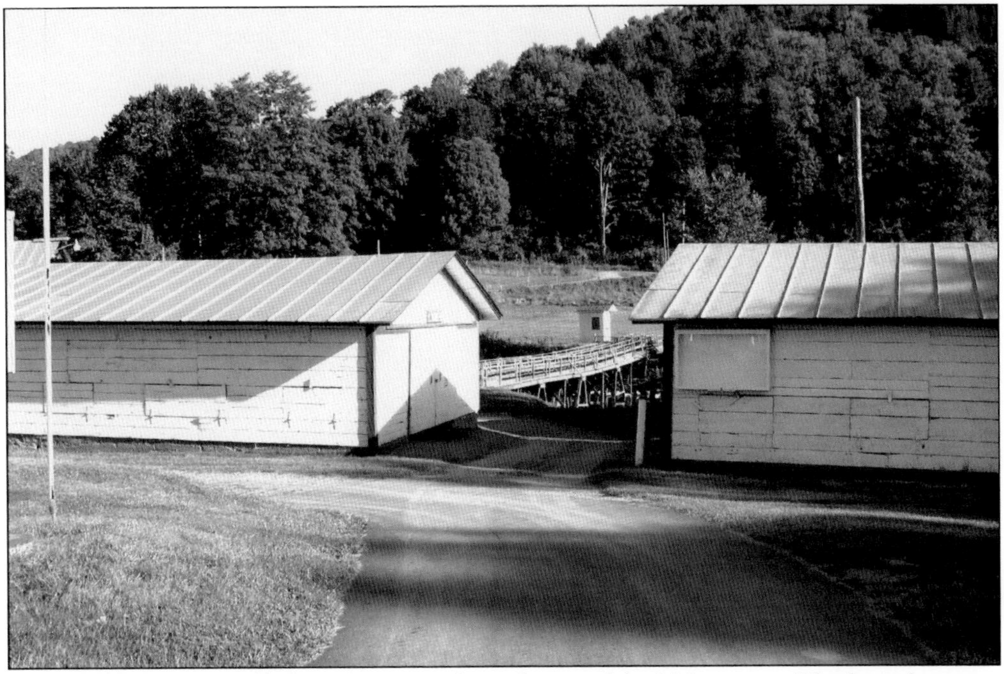

The cattle barns were added one-by-one in the early part of the 20th century. The fair is fortunate that all its barns are more than full at fair time. The footbridge (spanning the First Branch of the White River) is installed laboriously each summer and taken down after the fair. The ticket booth serves fairgoers who have parked in Sherlock's Field (north parking lot).

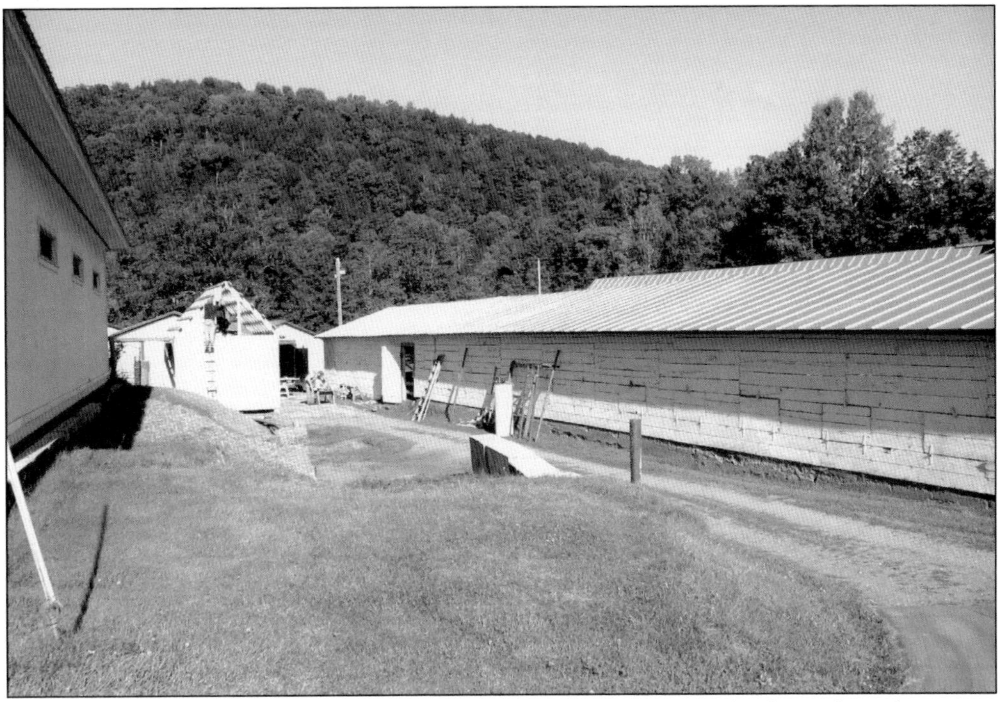

The milking parlor was built in 2007. Spectators can sit on the grassy bank watching dairy cows being milked.

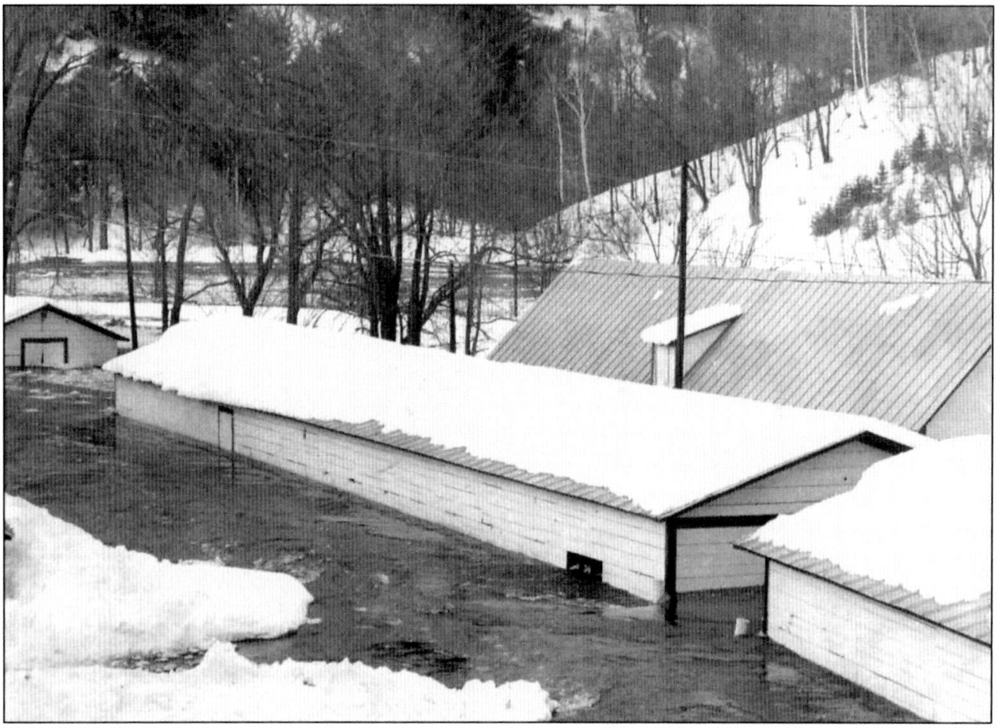

The main channel of the First Branch flowed between Floral Hall and the cattle barns in the spring of 1978. Fairgoers are impressed by signs painted on the barn sides showing those high-water marks.

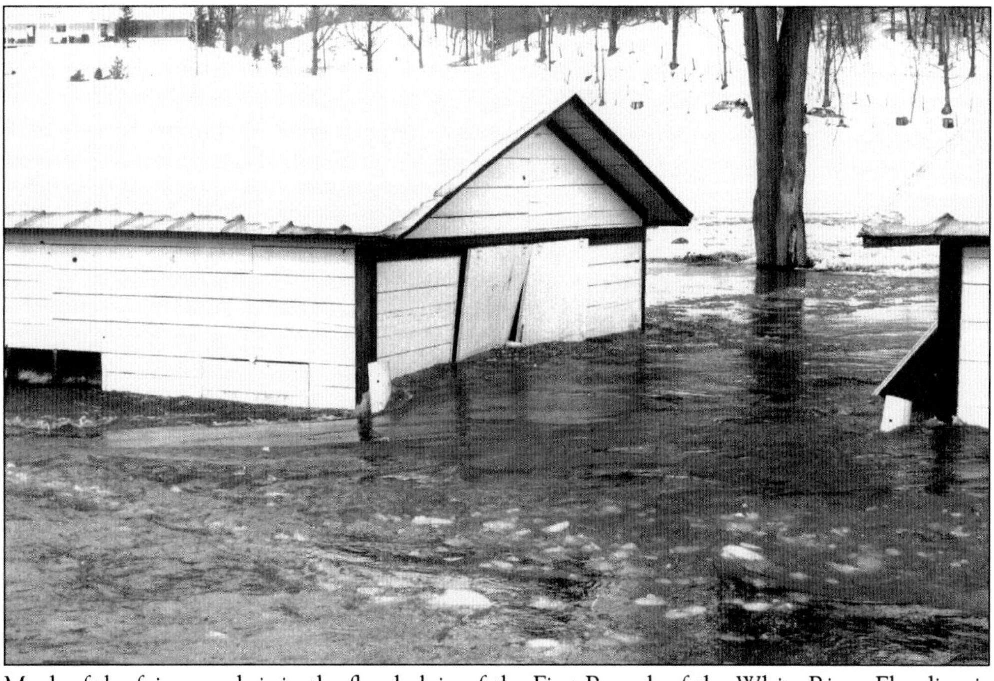

Much of the fairgrounds is in the flood plain of the First Branch of the White River. Flooding is caused either by an ice jam, heavy rains, or both at once. One of the worst disasters was in 1978.

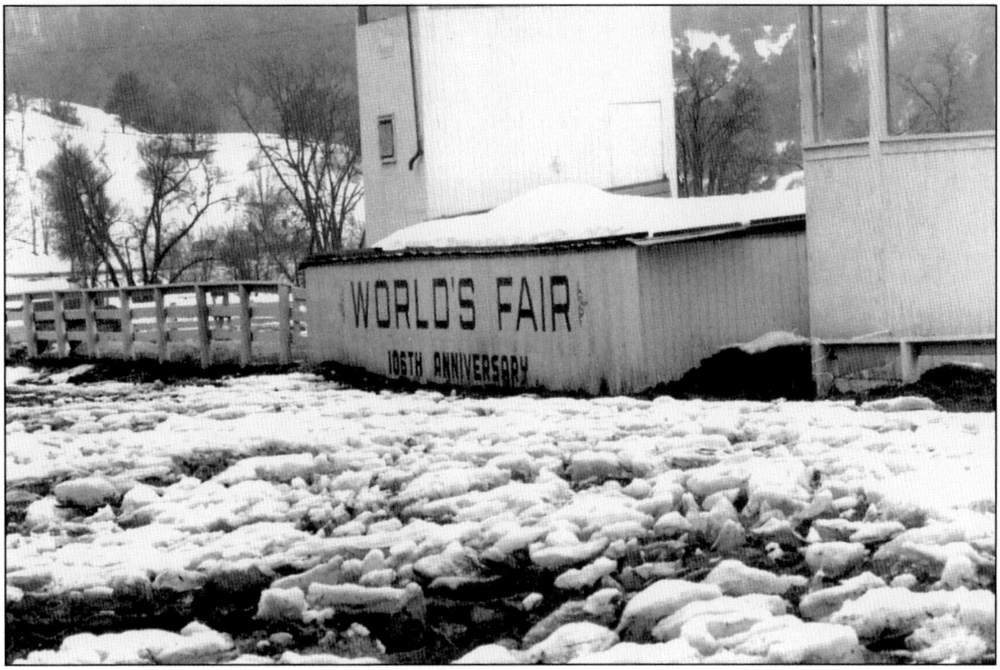

Following the 1978 flood the fair stage front sadly proclaimed "World's Fair 106th Anniversary." Amazingly most of the buildings suffered only minor damages.

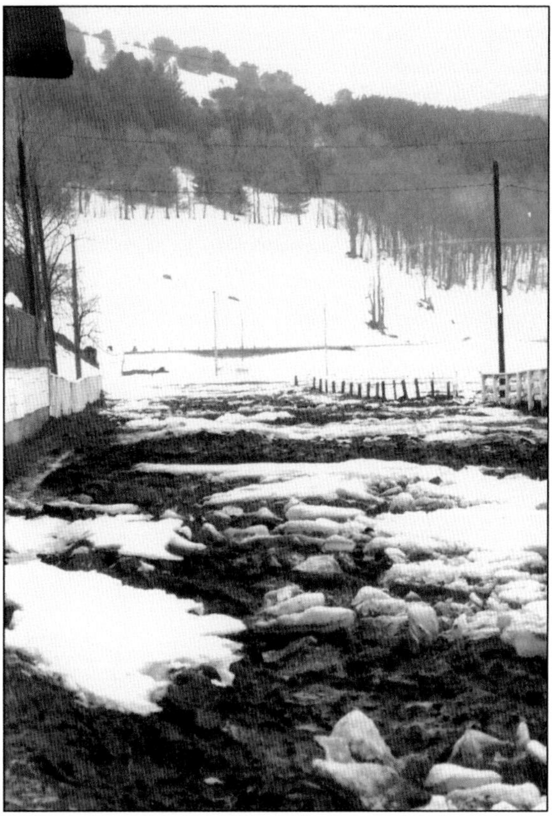

Looking south down the racetrack during the 1978 flood, only ice, mud, and debris could be seen. It is hard to imagine a fair-time sulky race there.

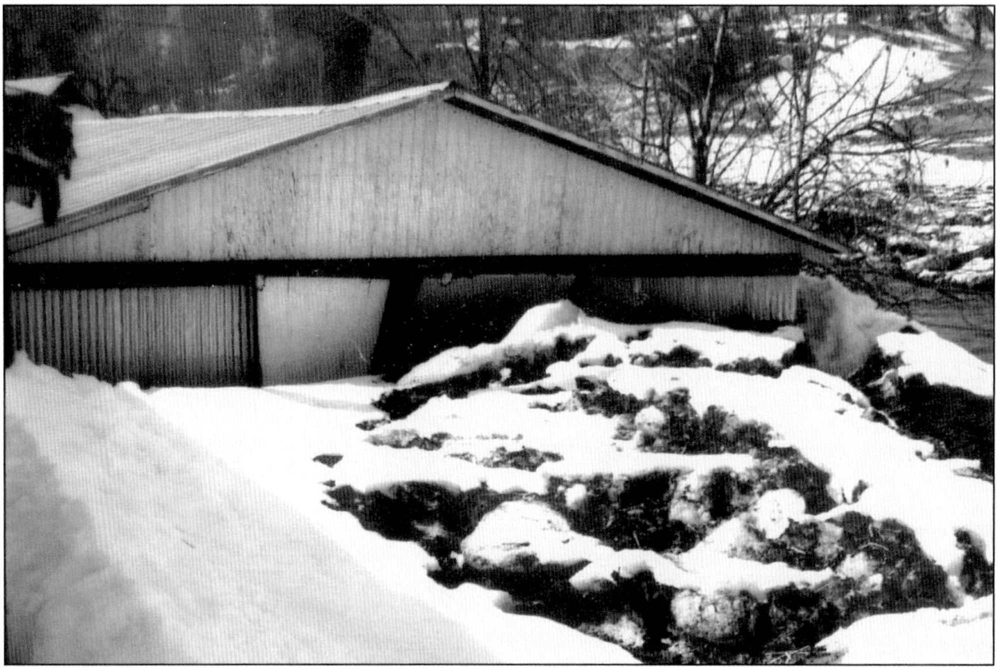

Damage to the barns has occurred several times over the years. The extensive clean-up necessary following such a flood is an enormous challenge for the grounds crew. It is amazing what they find that has washed downriver.

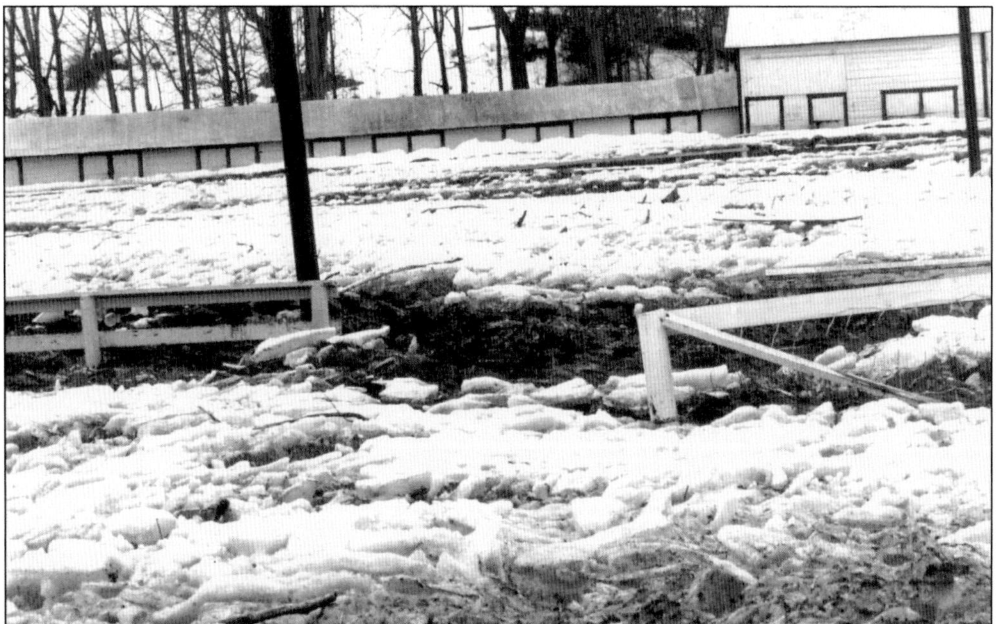

The center of the oval was a field of ice where the pavilion was built later. The barns were filled with ice to the ceilings, burying all stored farm equipment.

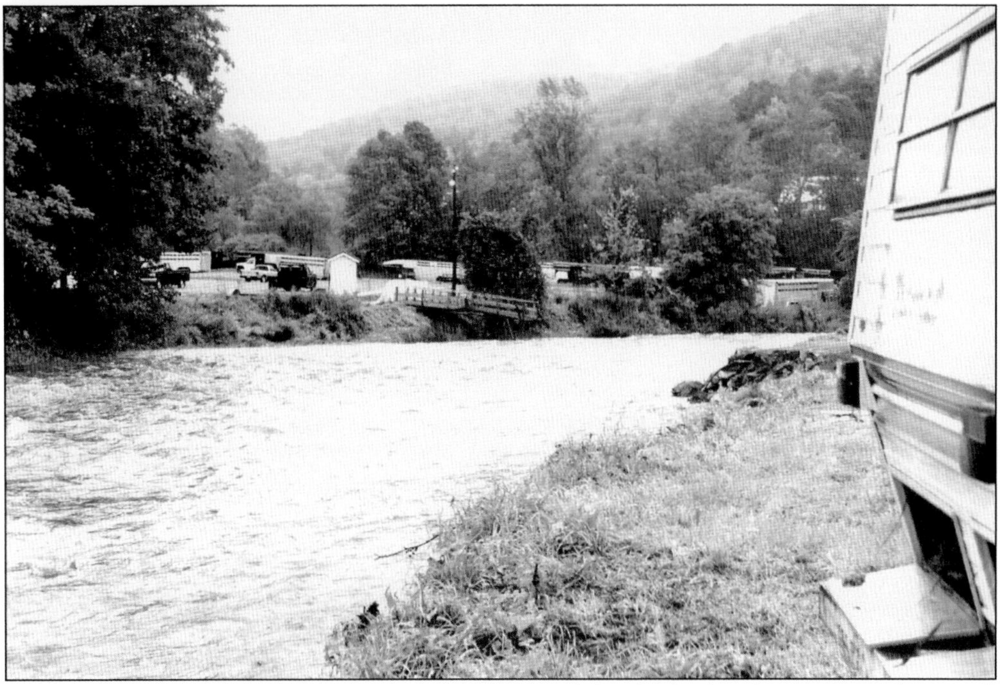

Hurricane Floyd, during the 1999 fair, washed most of the footbridge downriver. Only a small portion was left on the riverbank.

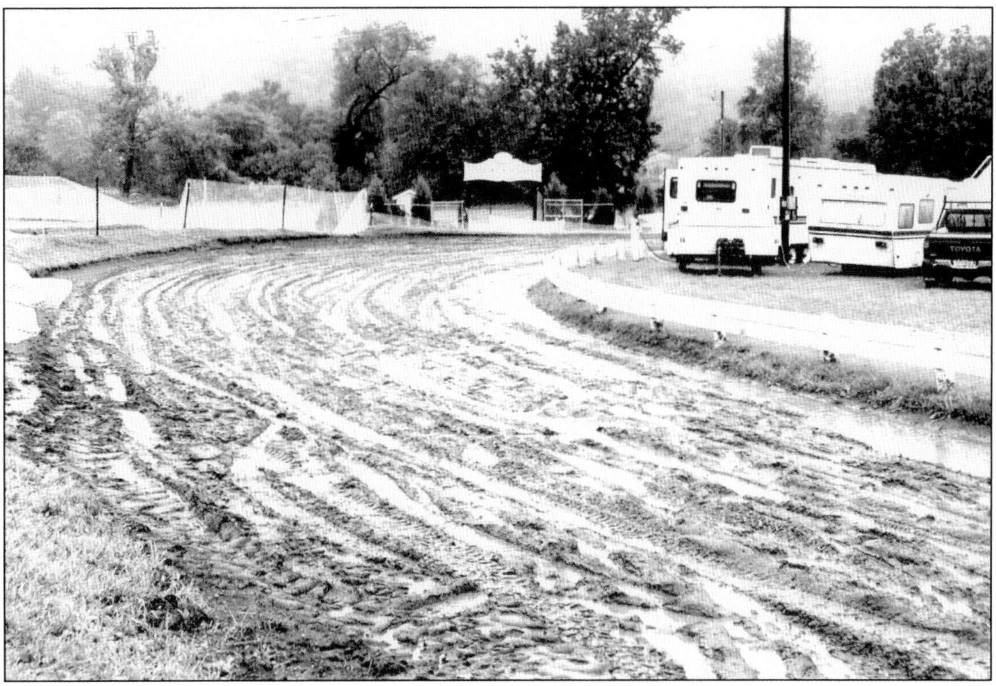

Hurricane Floyd left just mud where there had been a racetrack.

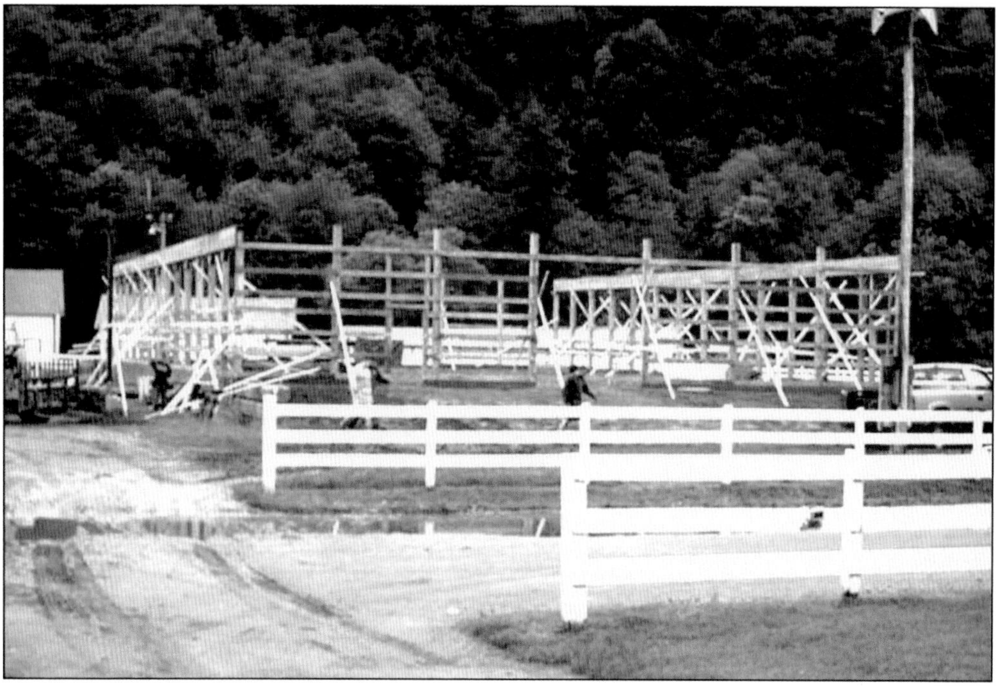

Tents were rented many years for commercial exhibits at fair time. The fair board of directors voted in 2005 to construct a pavilion for that purpose at the north end of the oval. The building is also used for many summer events.

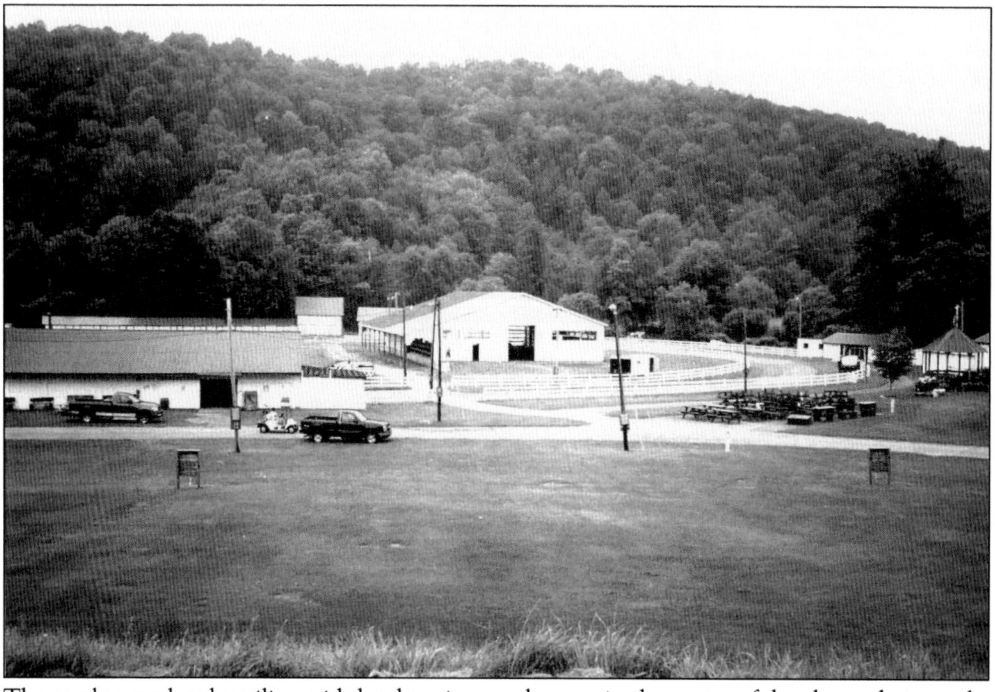

The newly completed pavilion with landscaping can be seen in the center of the above photograph.

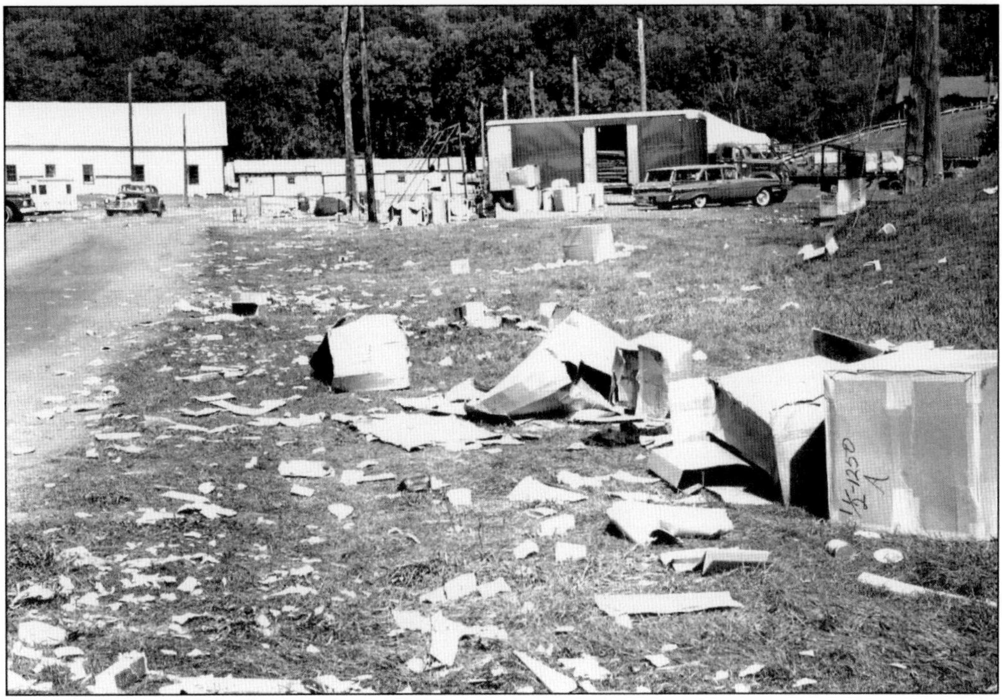

The midway area following the 1963 fair left a daunting task for the clean-up crew. The fairgrounds needed to be respectable before the heavy rains came. Much more garbage is generated nowadays; consequently, there are numerous trash containers all over the grounds, clean-ups are frequent at fair time, and recycling is making a big difference.

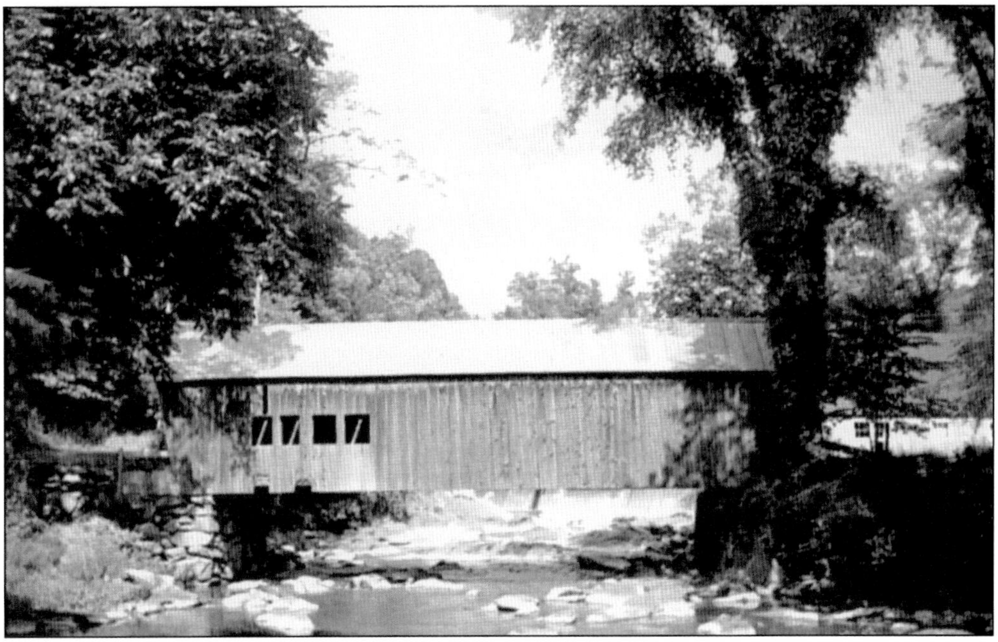

The Mill Covered Bridge in Tunbridge Village has served as the entrance to the fair's north parking lot since 1950. This is one of the very few fairs in the United States that provides the fairgoer such an opportunity.

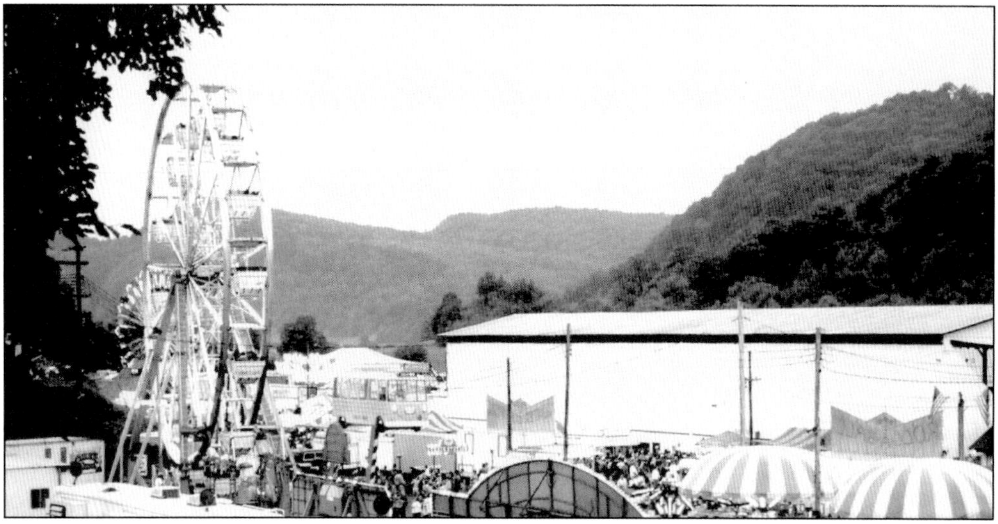

The Tunbridge fairgrounds is sandwiched between Vermont hills, the First Branch of the White River, and Tunbridge Village. It makes for a beautiful scene; however, the logistics of putting on the fair are made much more difficult. When dusk settles over the area, it is hard to imagine a more attractive fair location.

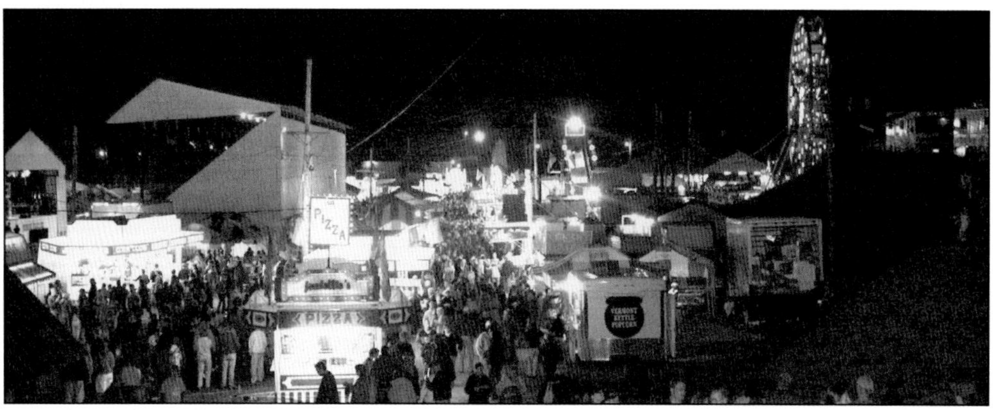

Saturday night at the fair is wall-to-wall with people enjoying the Tunbridge Fair's midway.

Two

ANTIQUE HILL

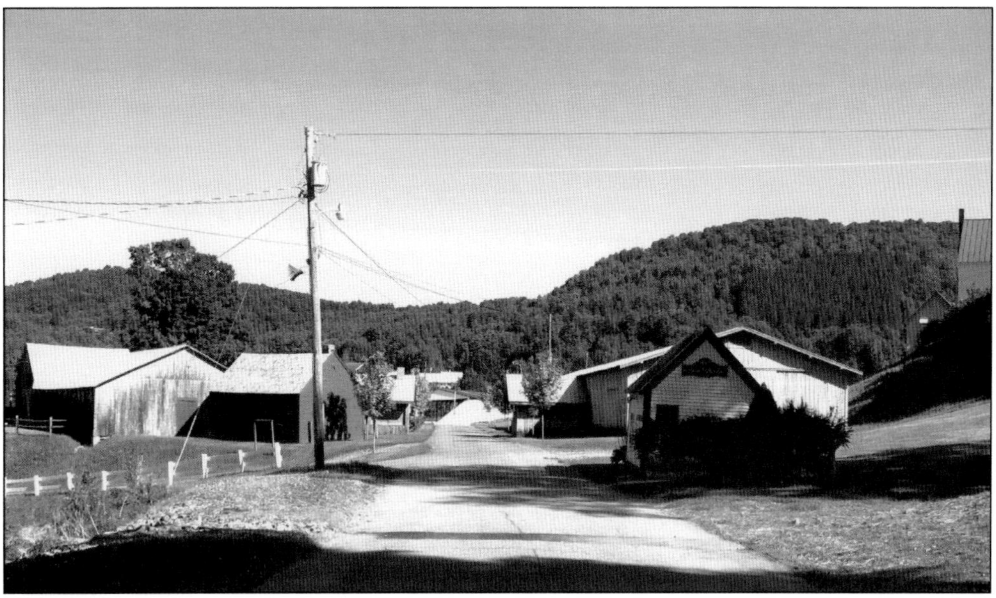

This photograph shows the south gate ticket booth on the right. A tree planting program has been placing maple trees on each side of the "main street" running through Antique Hill. The various antique buildings are on each side of the street.

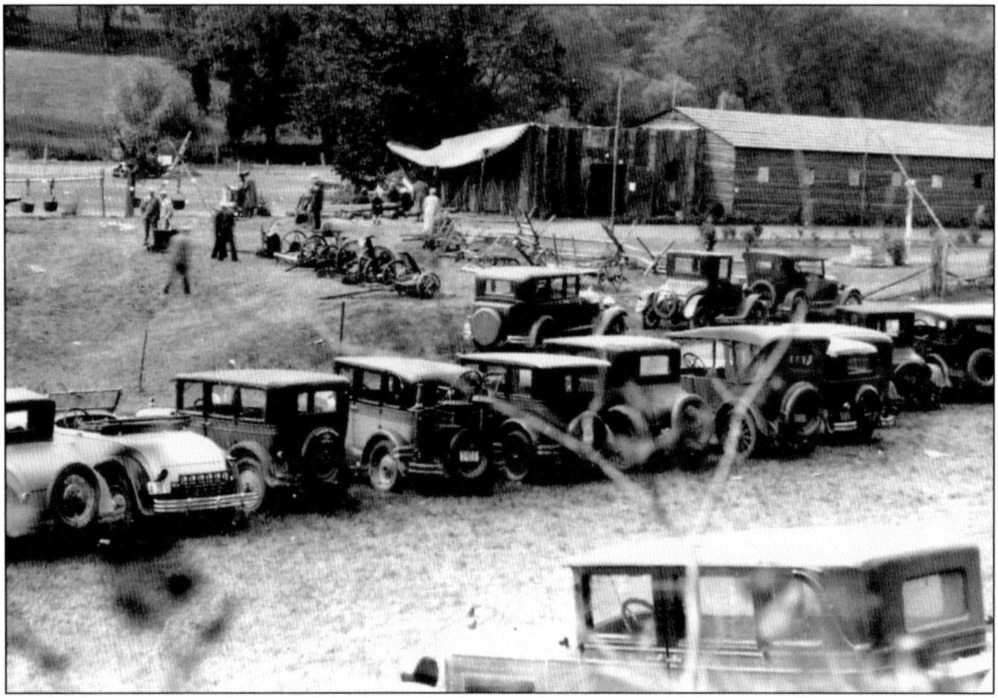

The first museum building on Antique Hill was established in 1929 under the direction of Edward Flint, the current fair secretary. For a number of years he had collected antiques and old farm equipment throughout the Tunbridge area. Many of the fair directors considered it all a waste of time and space; however, under Flint's supervision Antique Hill became an instant success.

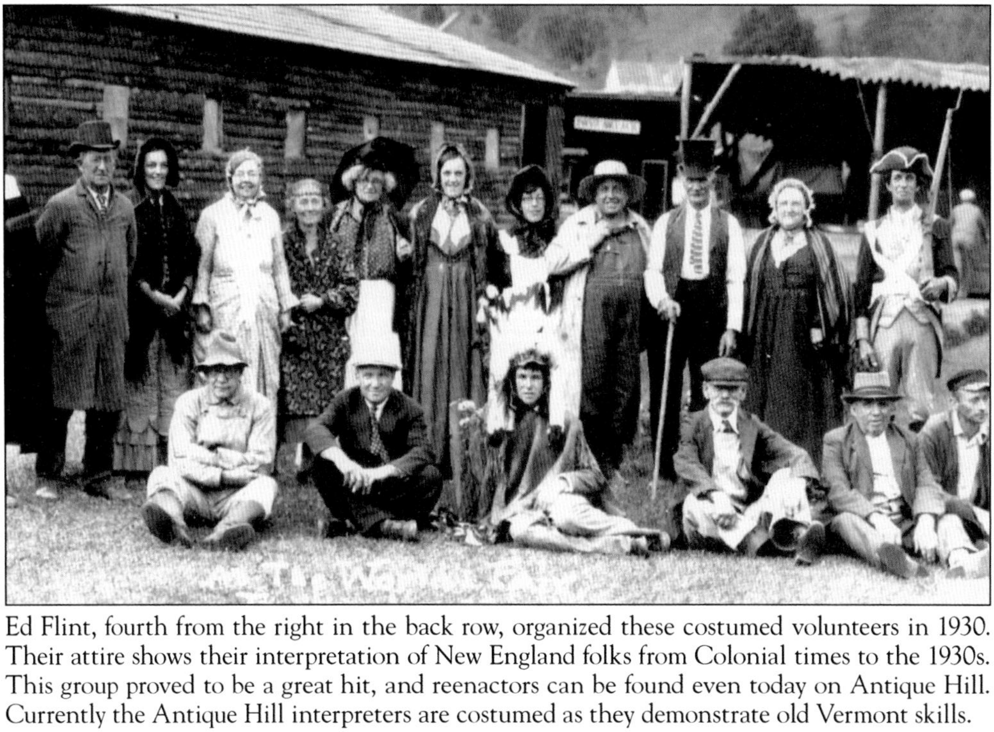

Ed Flint, fourth from the right in the back row, organized these costumed volunteers in 1930. Their attire shows their interpretation of New England folks from Colonial times to the 1930s. This group proved to be a great hit, and reenactors can be found even today on Antique Hill. Currently the Antique Hill interpreters are costumed as they demonstrate old Vermont skills.

Walter Haskins, portraying a Civil War soldier, and Ethan Jones, suggesting a soldier of the Revolution, stand guard in 1935.

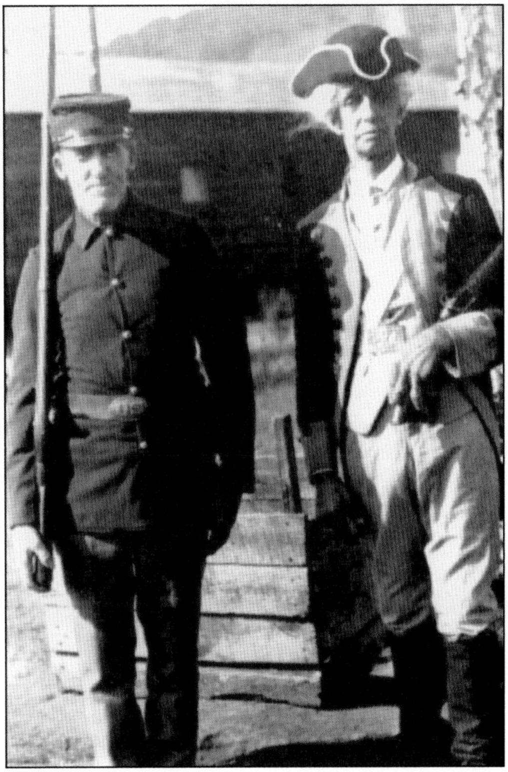

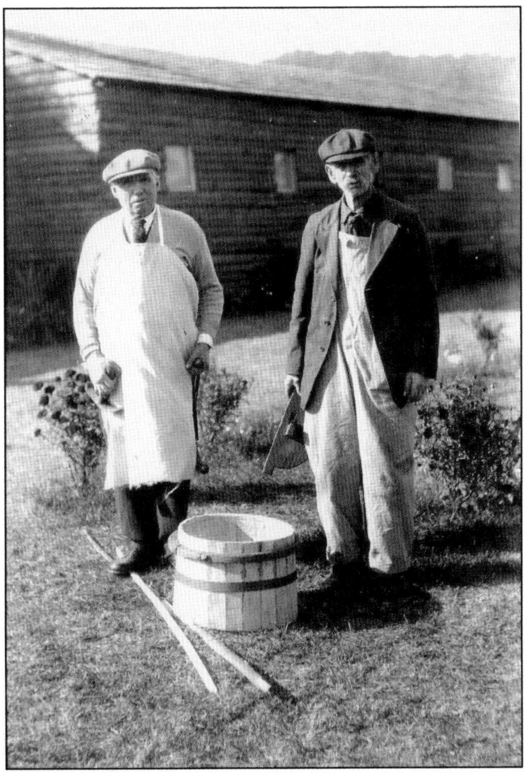

Cooper John W. Grant, on the right, and a helper in the 1930s pose by a barrel they have just finished.

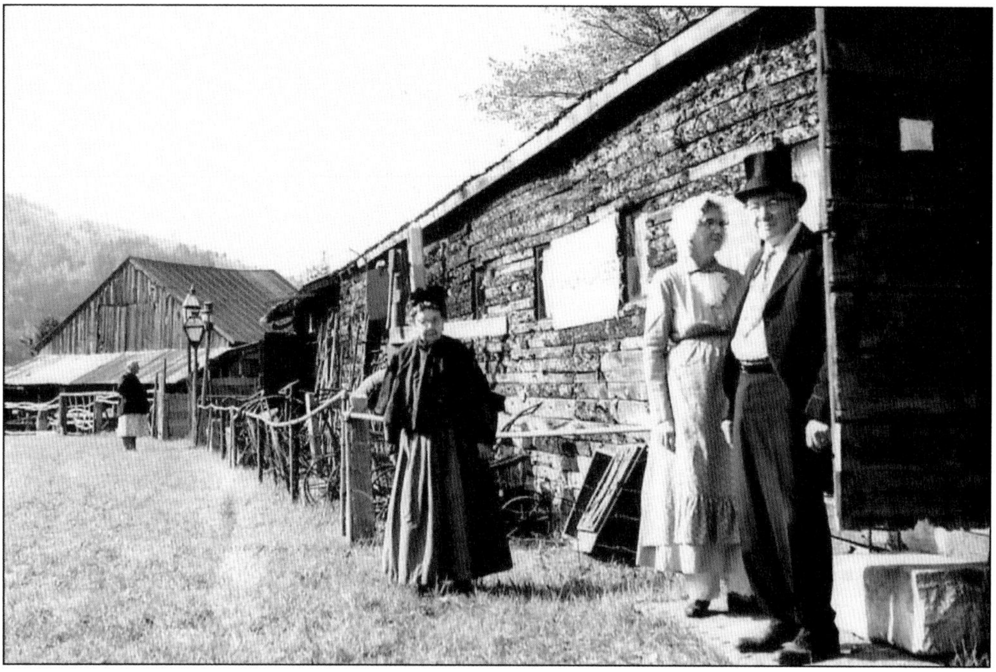

"Aunt Ruby" Pierce, Stella Waldo, and Walter Sawyer pose in 1930 in front of the Log Cabin Museum. Aunt Ruby was a fixture on Antique Hill for about 30 years. She also made fair commercials for the radio, singing a Tunbridge Fair song (of her own composition) to the tune of "Jingle Bells." The rendition was "truly remarkable."

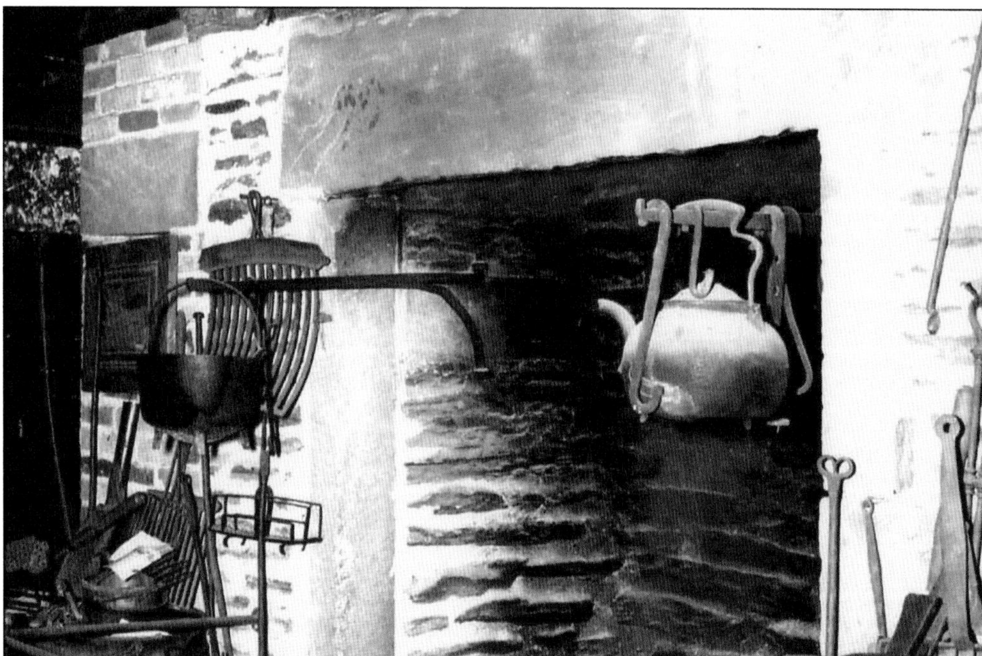

The original antique cabin was constructed in 1929. The outstanding feature was the fireplace using local stones. When that building was replaced with a larger museum in 1972, under the direction of Carol Möller, the only things reused were the fireplace and chimney.

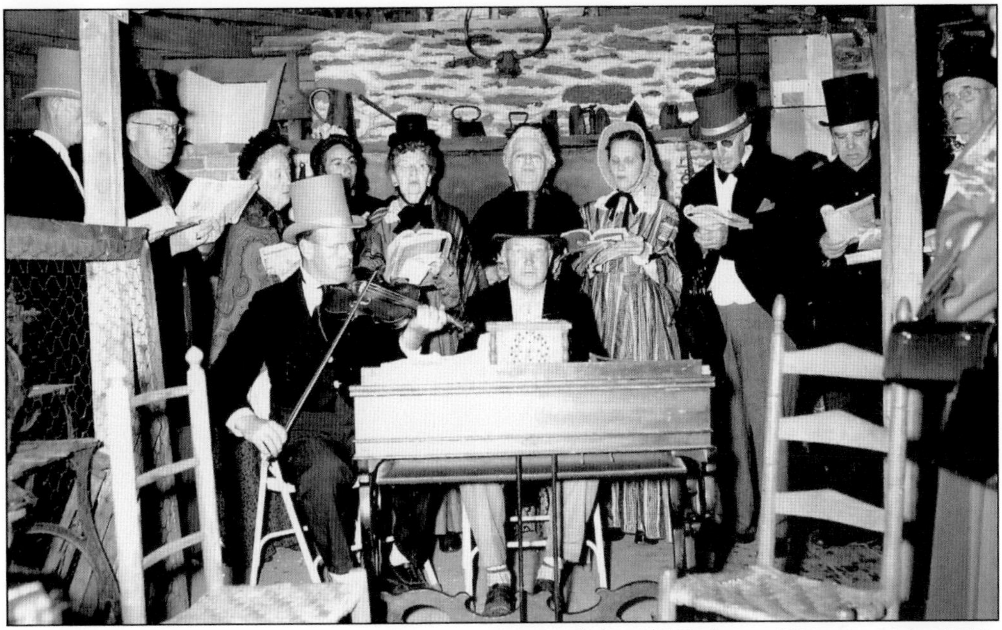

Charles (Chuck) Luce and Walter Sawyer accompanied the Old Fashioned Singers in the 1940s into the 1960s. The group was organized by Ruby Pierce, fifth from the left in the back row. Usually the group performed outside; therefore, it can be assumed it was raining.

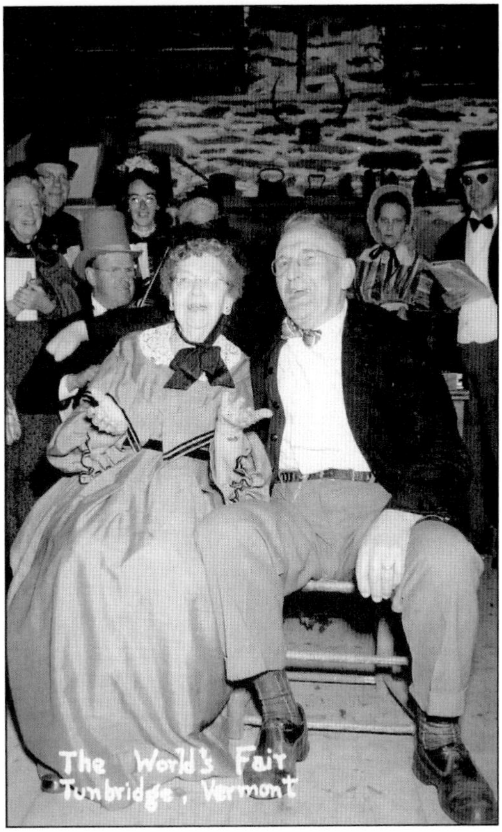

Ruby Pierce and an admirer are seen enjoying themselves (backed by the Old Fashioned Singers). For many years Pierce was the cabin hostess. The dress and hat she wore are displayed in the museum.

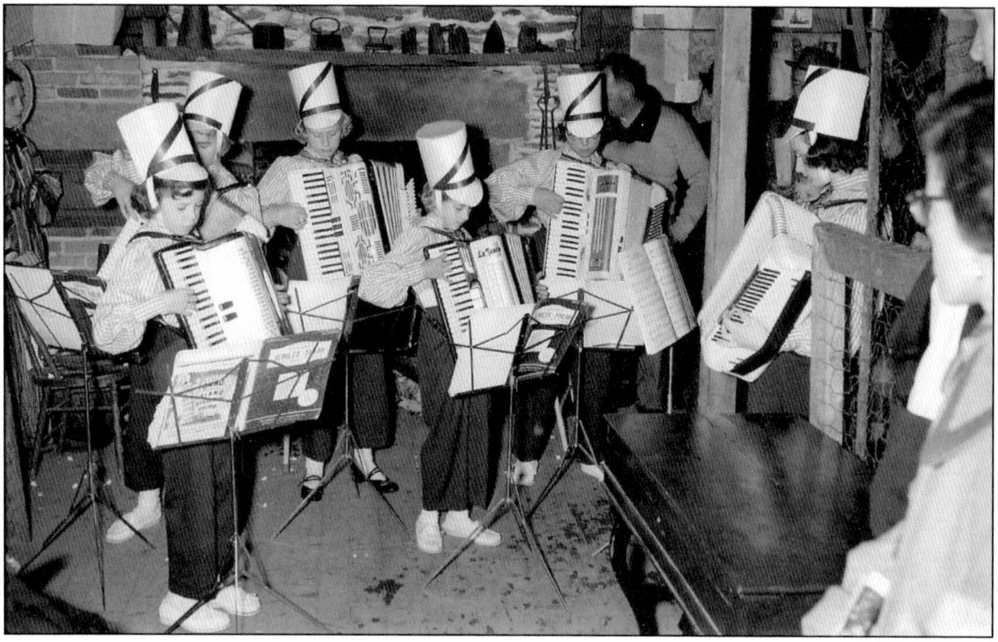

The Barre Accordion Band played at several fairs during the late 1930s. They were playing in this photograph inside the cabin museum; so they were probably unable to be on the main stage because of inclement weather.

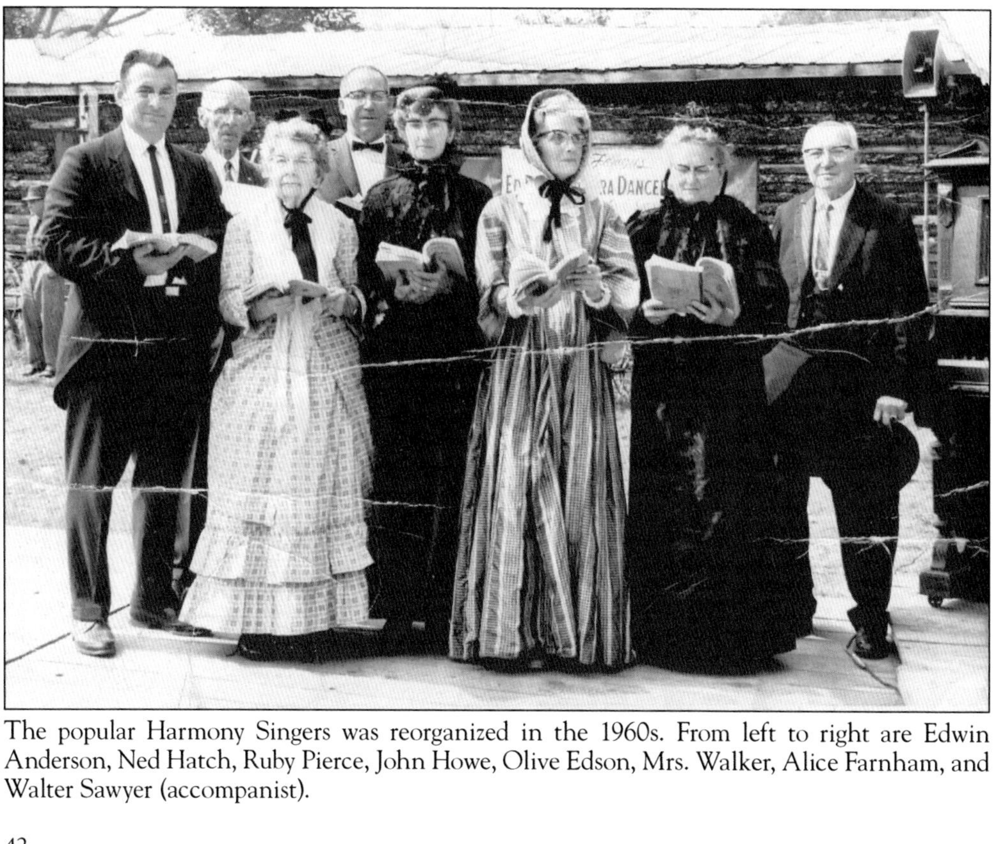

The popular Harmony Singers was reorganized in the 1960s. From left to right are Edwin Anderson, Ned Hatch, Ruby Pierce, John Howe, Olive Edson, Mrs. Walker, Alice Farnham, and Walter Sawyer (accompanist).

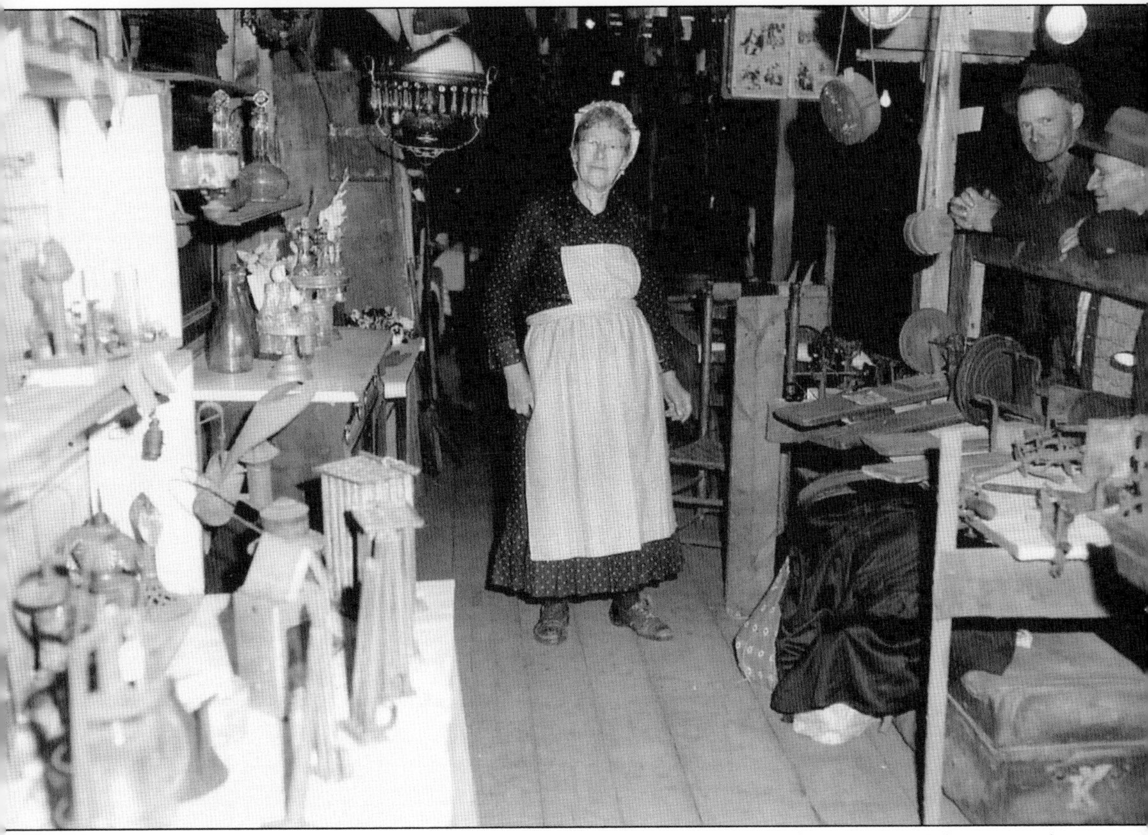

A proud volunteer poses in 1941 amid an extremely jumbled display in the museum. The original cabin proved to be far too dark and small for the growing collection; consequently, aisles were narrow and items were cluttered, thus preventing enjoyment of the antiques.

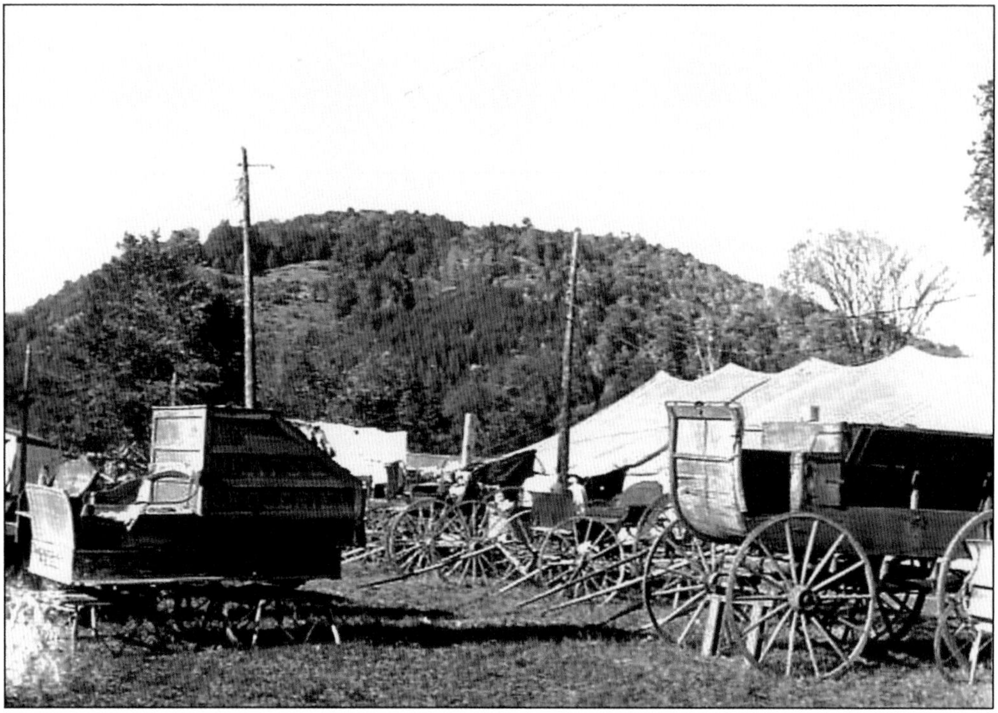

In the early years the fair could not afford additional buildings on the hill. The horse-drawn vehicles and farm equipment had to be hauled out yearly onto the green for display.

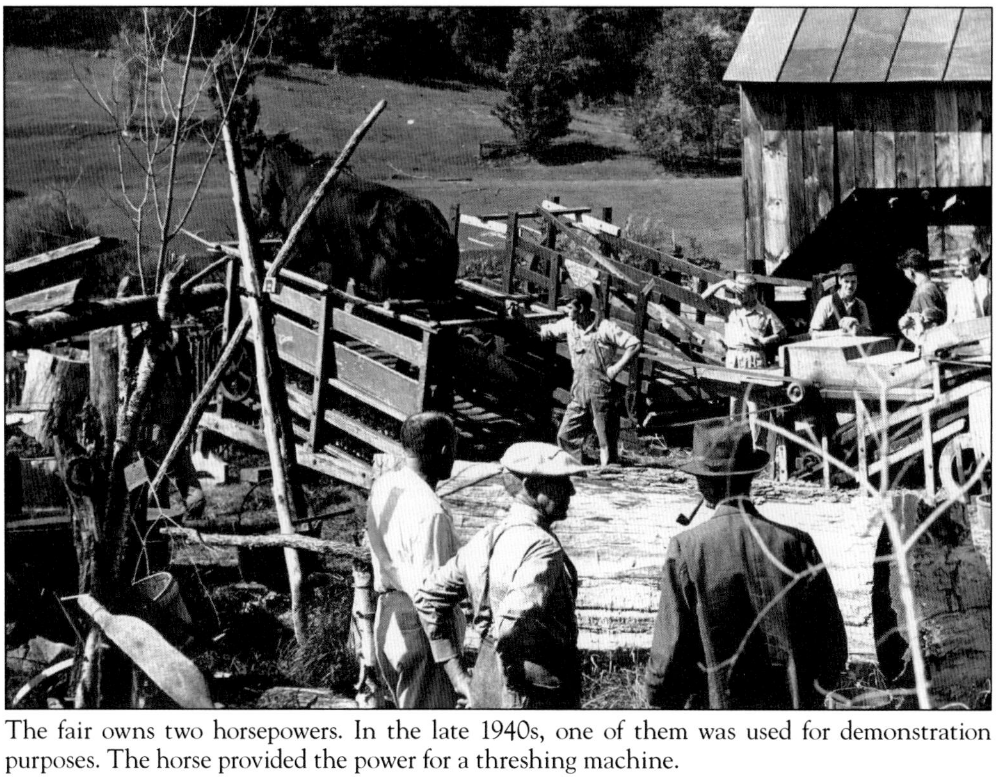

The fair owns two horsepowers. In the late 1940s, one of them was used for demonstration purposes. The horse provided the power for a threshing machine.

These two well-clad gentlemen perhaps were discussing the merits of a fair sleigh. The impressive transportation collection of horse-drawn vehicles gives modern-day visitors a good idea of what traveling was like a century ago.

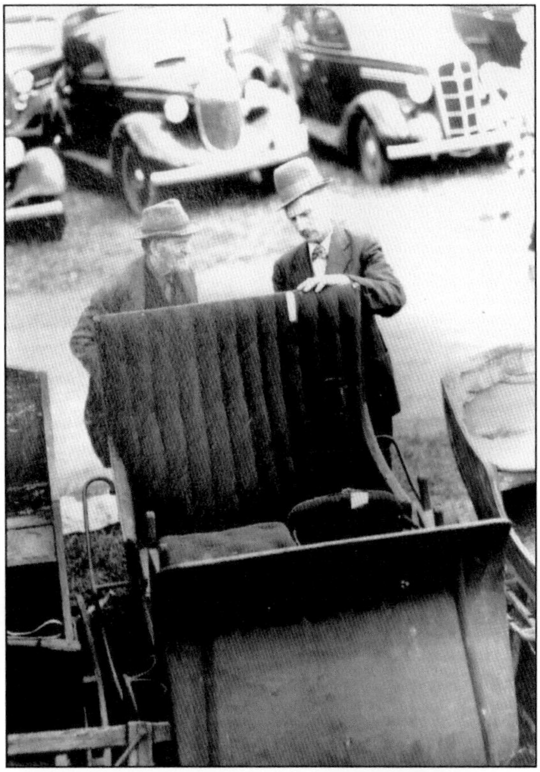

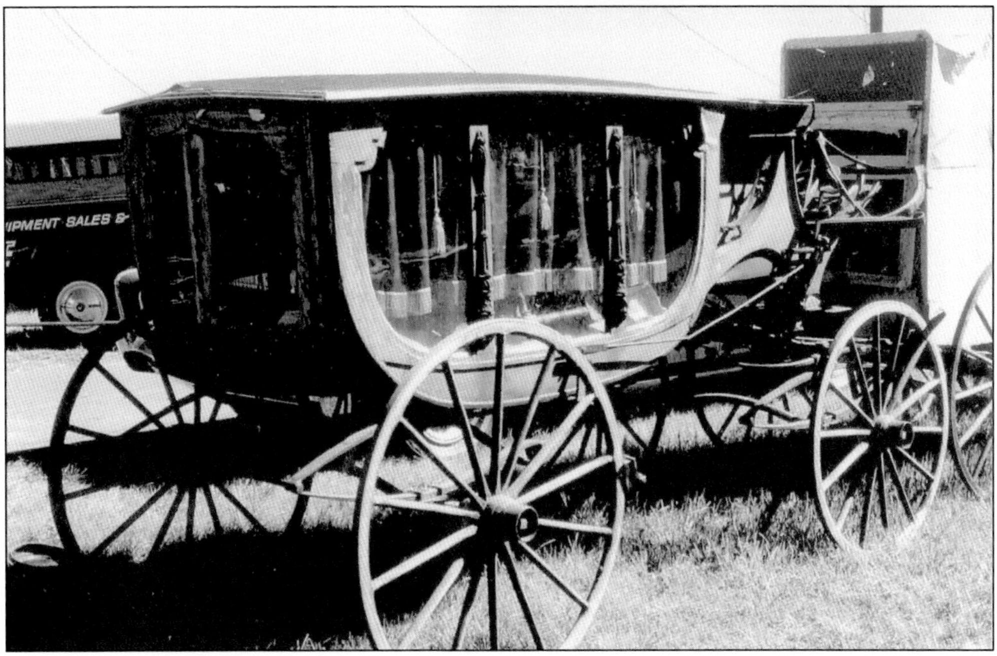

The town of Tunbridge purchased a hearse in 1886 for $335.50. The hearse driver for the next 30 years transported bodies of most Tunbridge residents to their final resting places. Edward Flint procured it in 1929 for the fair's museum. The collection also includes the Chelsea hearse with its winter runners.

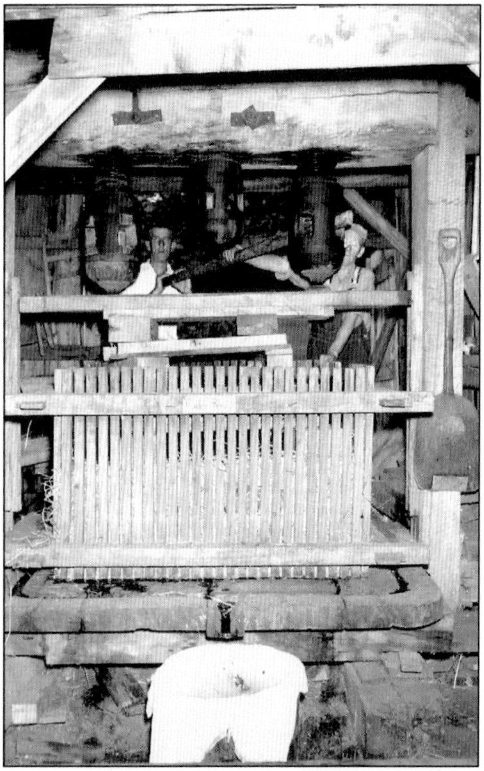

Jesse Wood of Strafford made and used this huge, one-of-a-kind cider press in 1835. It features handmade wooden screws that are six inches in diameter by eight feet long. Now it is displayed in its own shed next to the Log Cabin Museum.

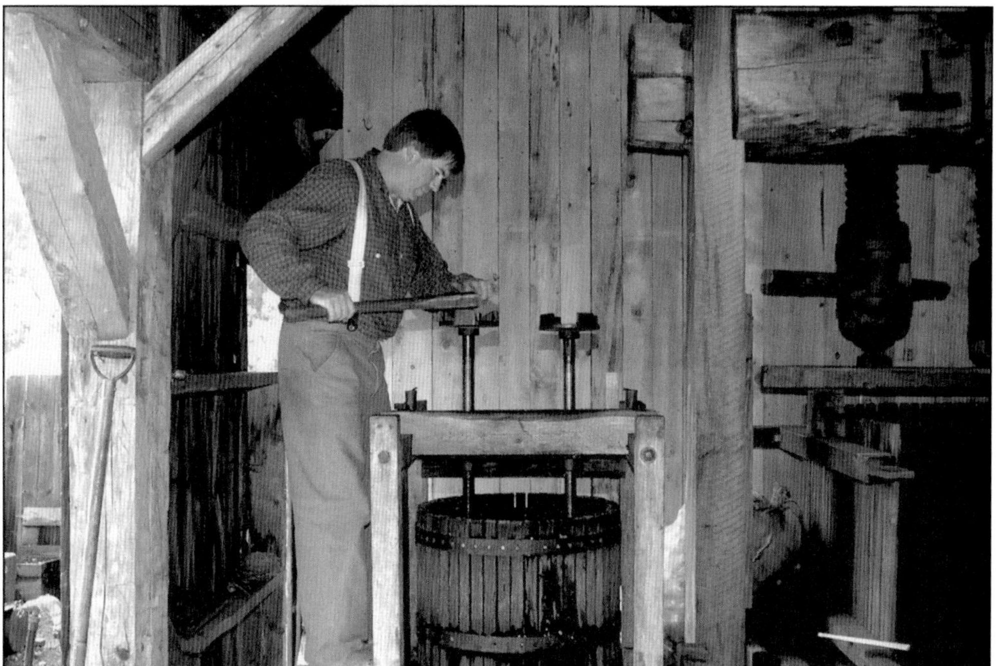

Dennis Cilley, the current superintendent of antique vehicles and farm equipment, works the small cider press. Hundreds of cups of apple cider are given away each fair time. (Courtesy of Mark Dixon.)

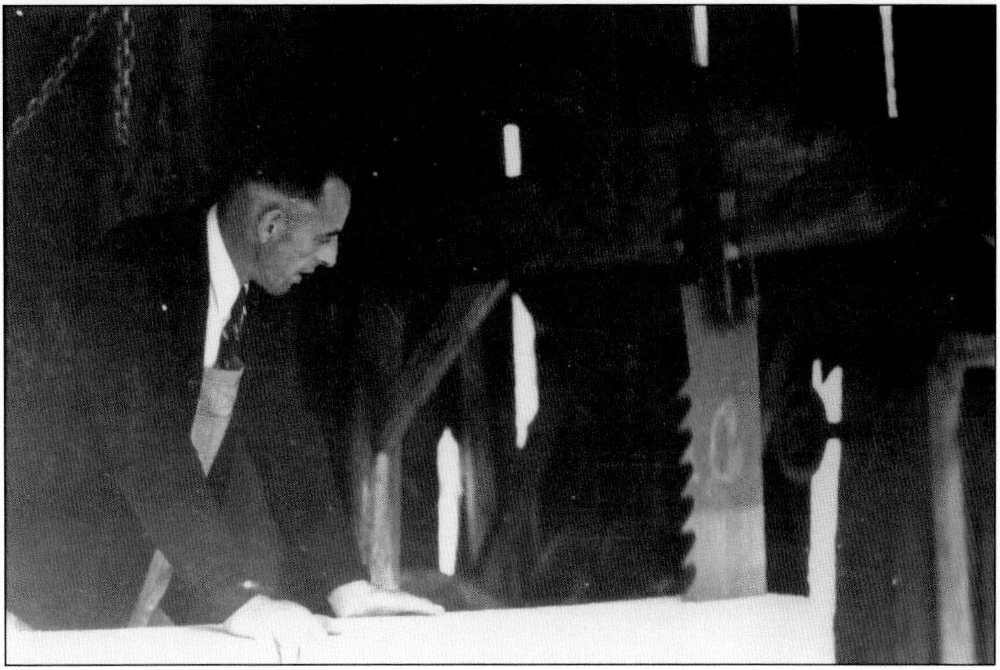

Edward Flint rescued this up-and-down sawmill from the old Shepard Mill in Royalton. Earl Larkin operated it in the 1930s and 1940s. It was sold to the Shelburne Museum in 1951.

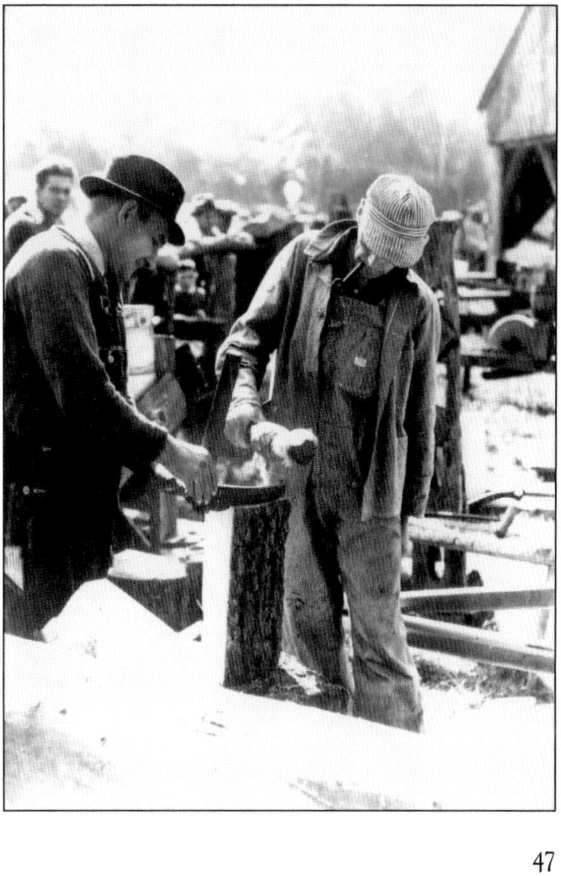

Shingle making using authentic hand tools was, and still is, demonstrated on the hill. The shingle maker, right, enlisted a fairgoer as his assistant. Shingle making was an important industry in Tunbridge for decades by local mills, which produced hundreds of thousands yearly.

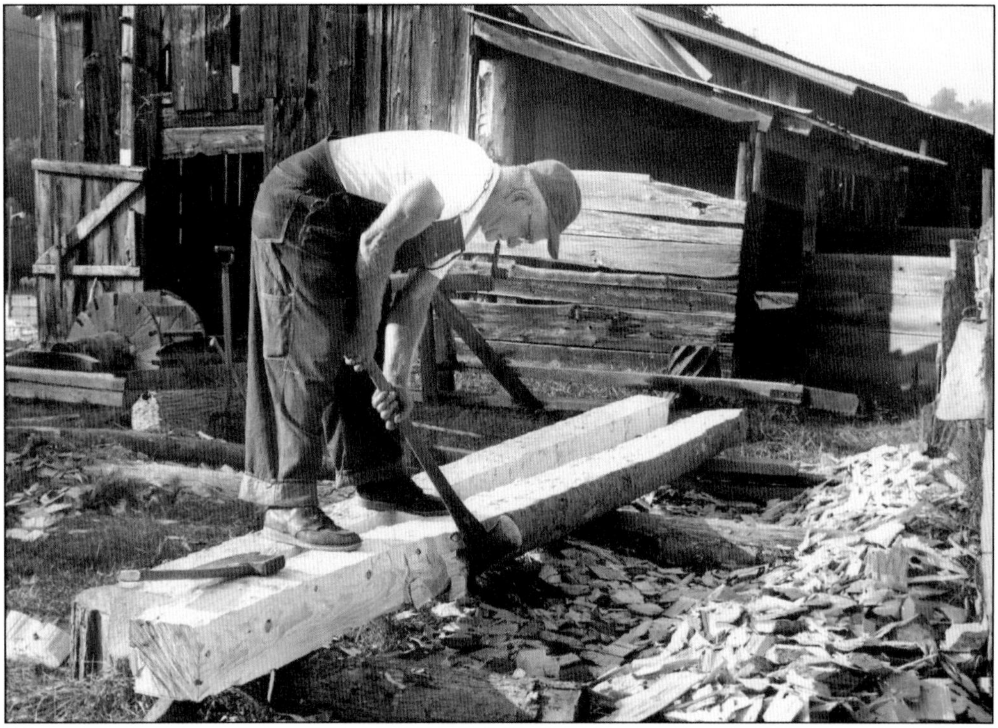

Les Shirlock of Royalton hewed logs for many years at fair time. The appearance of his area seems to indicate that it was probably near the end of that 1950s fair.

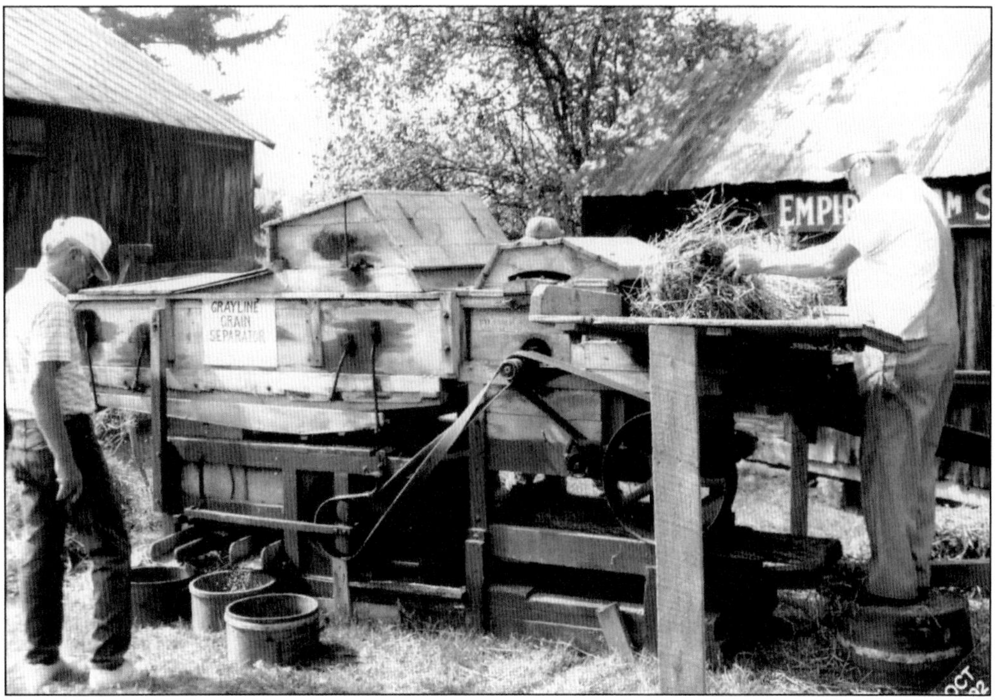

In this photograph, Sid "W. S." Gilman, right, works the Grayline Grain Separator. Gilman served as superintendent of outside antiques from 1980 to 1995.

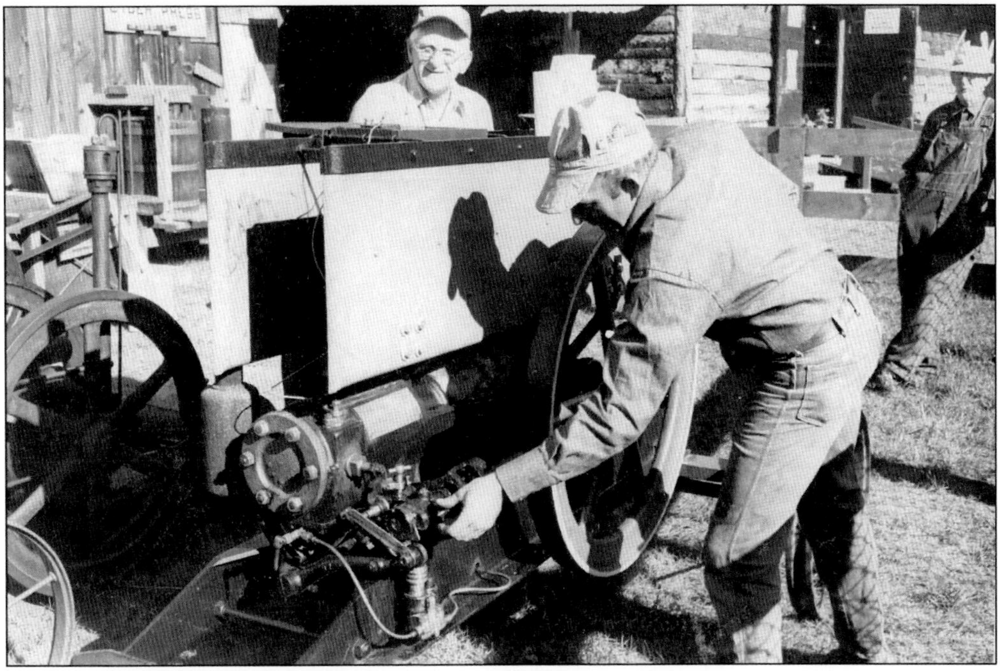

W. S. Gilman, left, and Hale Mattoon, right, prepare to "fire up" the fair's Abenaque gas engine during a 1960s fair. The Bellows Falls–built engine is the largest in the collection. It is still demonstrated yearly.

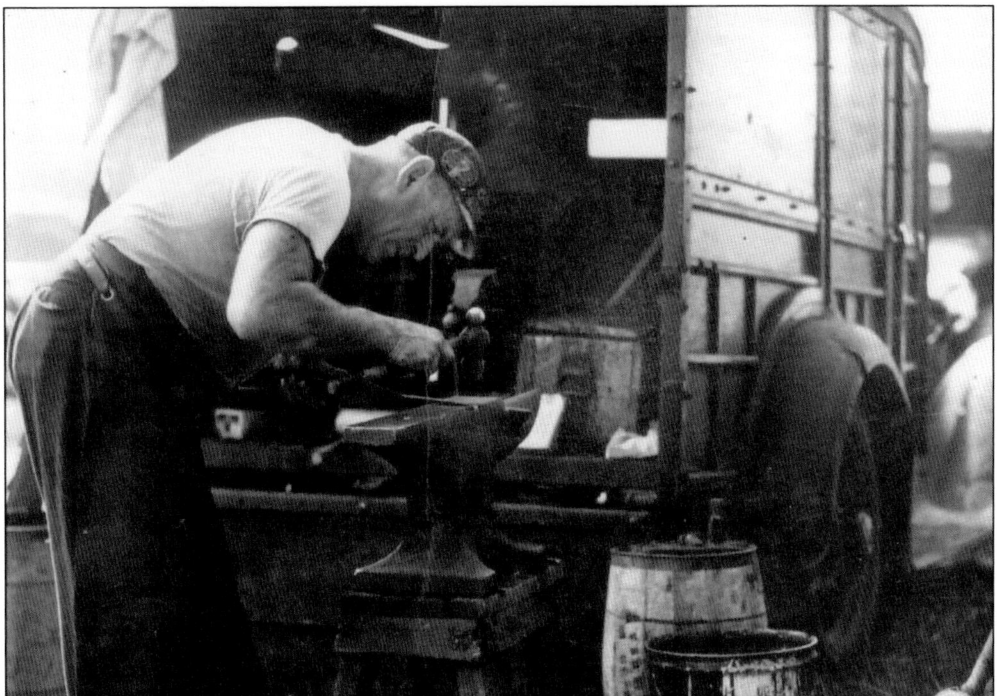

In the 1930s and 1940s a "traveling" blacksmith was on duty. His mobile shop contained everything he needed for shoeing animals and other related jobs.

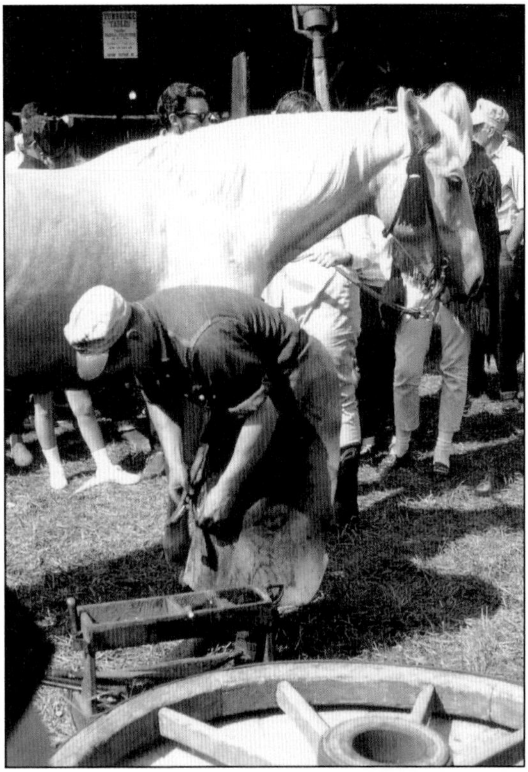

During the mid-1960s a farrier shoed horses on the green in front of the museum. That was discontinued because of liability. The poster tacked on the museum is for Tunbridge tables made by John Rowell.

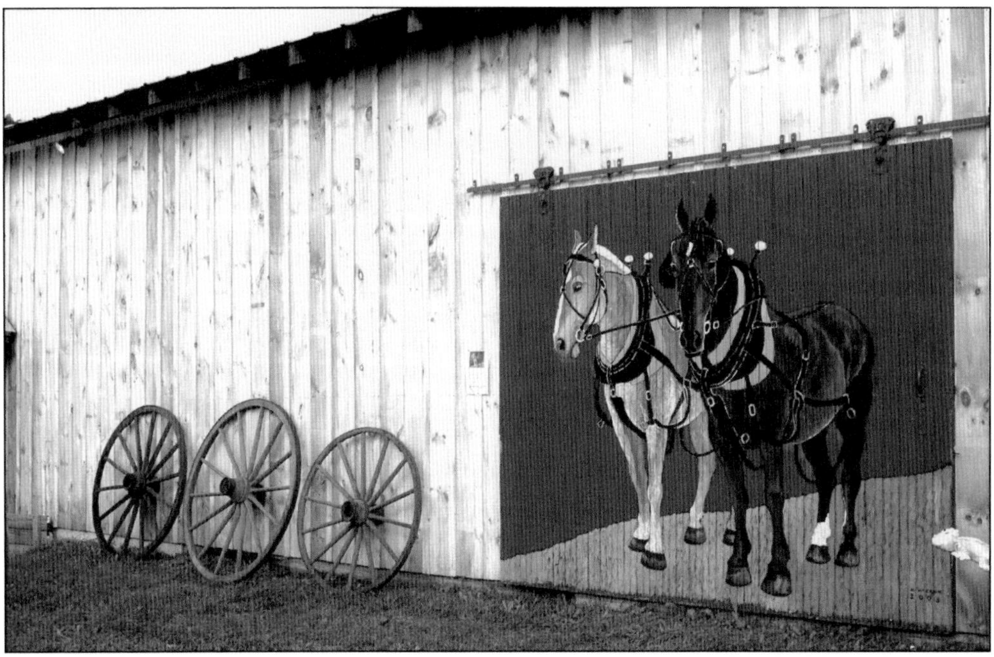

The 1960s metal exhibition building got wood siding in 2006. It now houses horse-drawn vehicles and farm equipment. One of the sliding doors has an original painting of two workhorses by local artist Marianne McCann. Bill and Marie Danforth loan it yearly to the fair. (Photograph by Mark Dixon.)

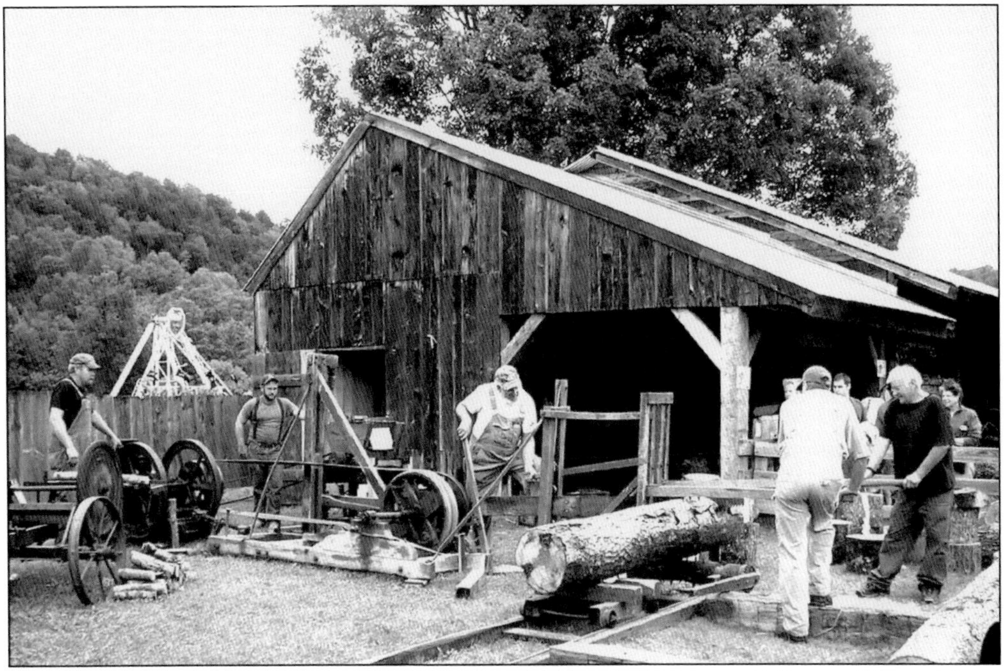

From left to right, Whitey Adams, Shane Young, Bill Danforth, Dennis Cilley, and Elliott Morse demonstrate the dragsaw powered by a gas engine. Throughout the four days of the fair these men show how various pieces of antique farm equipment were used about 100 years ago.

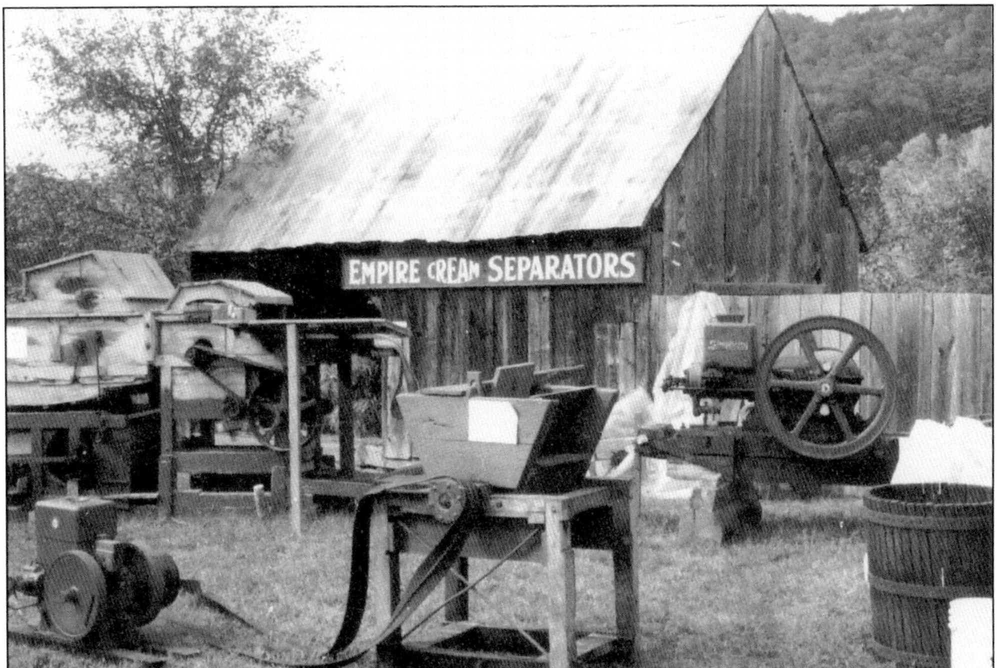

An excellent collection of gas engines of various sizes is used during the fair to power some of the farm equipment. Fairgoers enjoy hearing the "pop, pop" sounds of the antique engines.

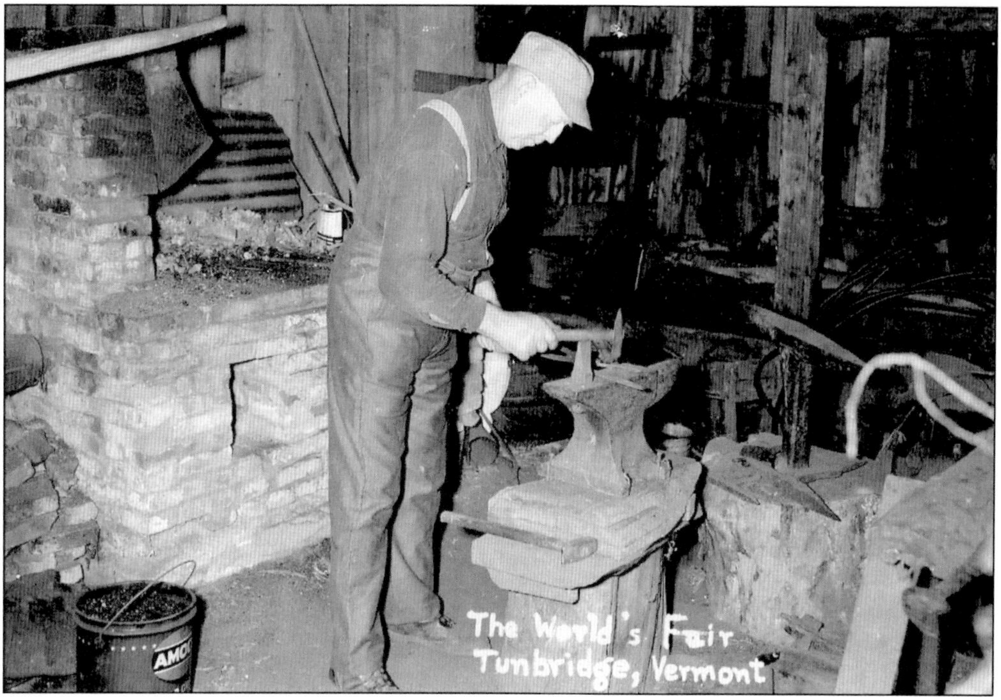

The original Log Cabin Museum located a blacksmith's area at its south end. When he fired up the forge the consequent smoke and noise did not create a good situation. The blacksmiths are now housed in their own shop building. Andrew Nielson did blacksmithing in the 1940s and 1950s.

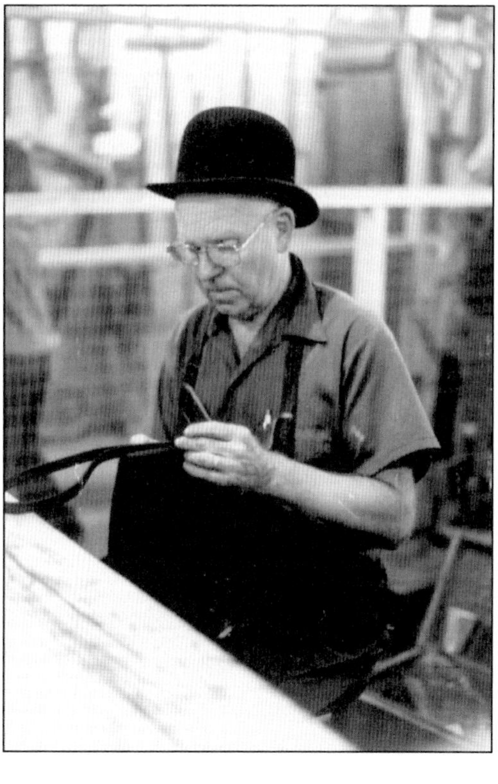

The Sanborn family of Chelsea ran a harness shop for several generations. Glen Sanborn demonstrated his harness making skills for years in the museum.

Howard Stockwell wove beautifully intricate designs on caned chairs for a few years. (Courtesy of Arlene Stockwell.)

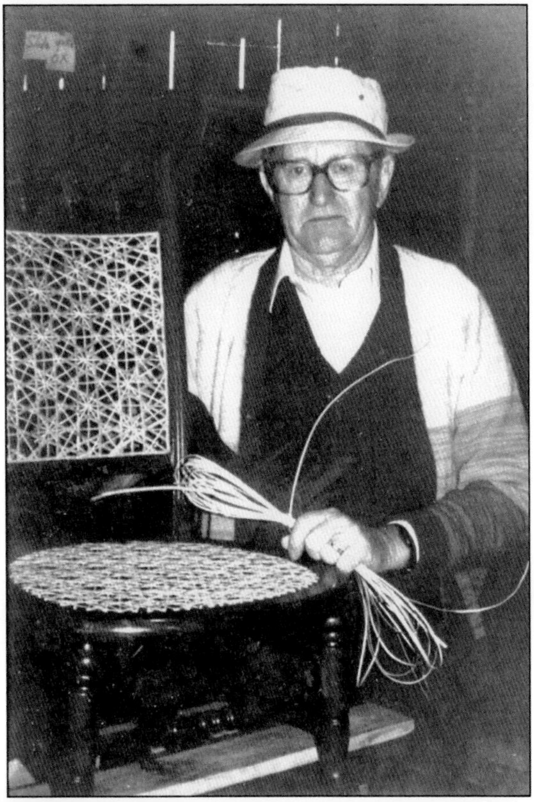

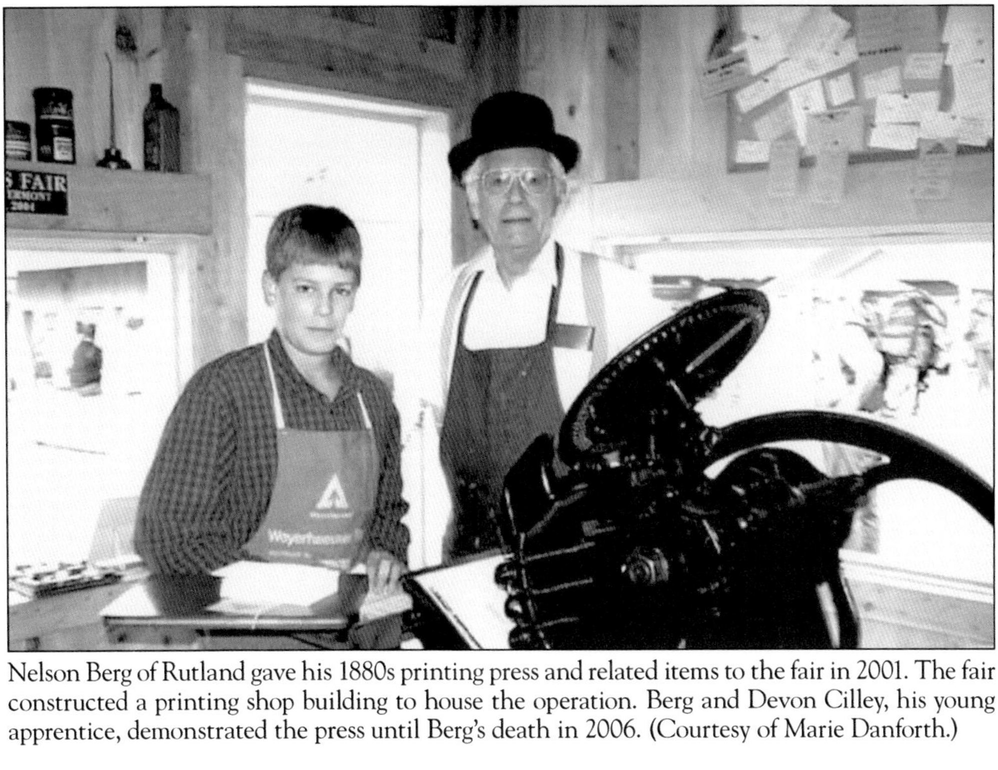

Nelson Berg of Rutland gave his 1880s printing press and related items to the fair in 2001. The fair constructed a printing shop building to house the operation. Berg and Devon Cilley, his young apprentice, demonstrated the press until Berg's death in 2006. (Courtesy of Marie Danforth.)

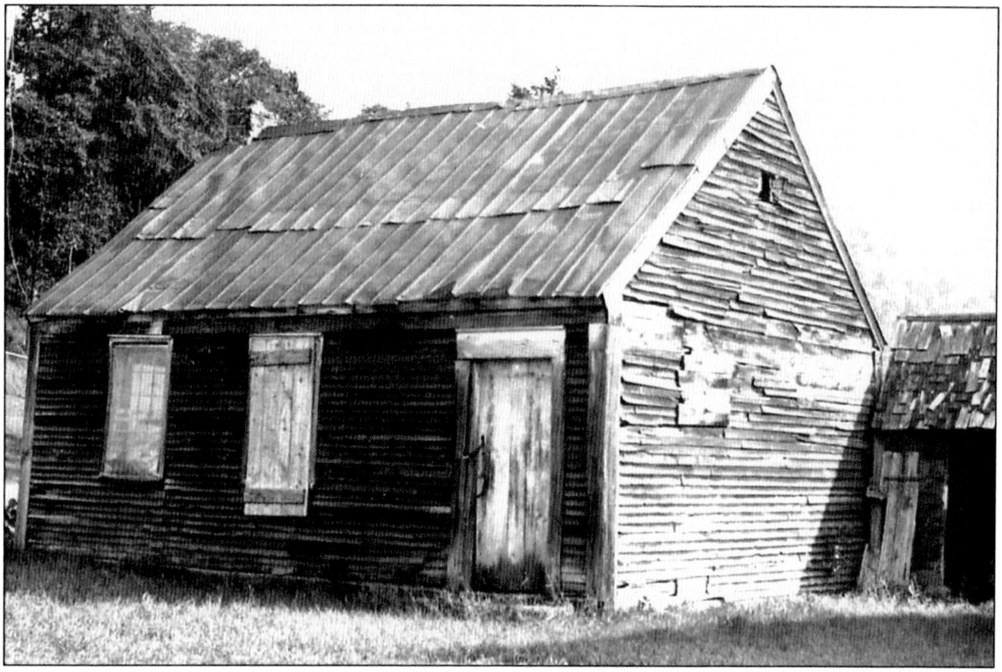

The District No. 8 1840 schoolhouse and its privy, right, were originally located at the corner of Bicknell Hill and Larkin Roads. This very rural school was closed by 1905. Edward Flint in 1940 hired Rudolph "Shorty" Danforth to move the structure about six miles to the fairgrounds. Until it was restored in the 1970s the building was displayed in the shabby condition seen above.

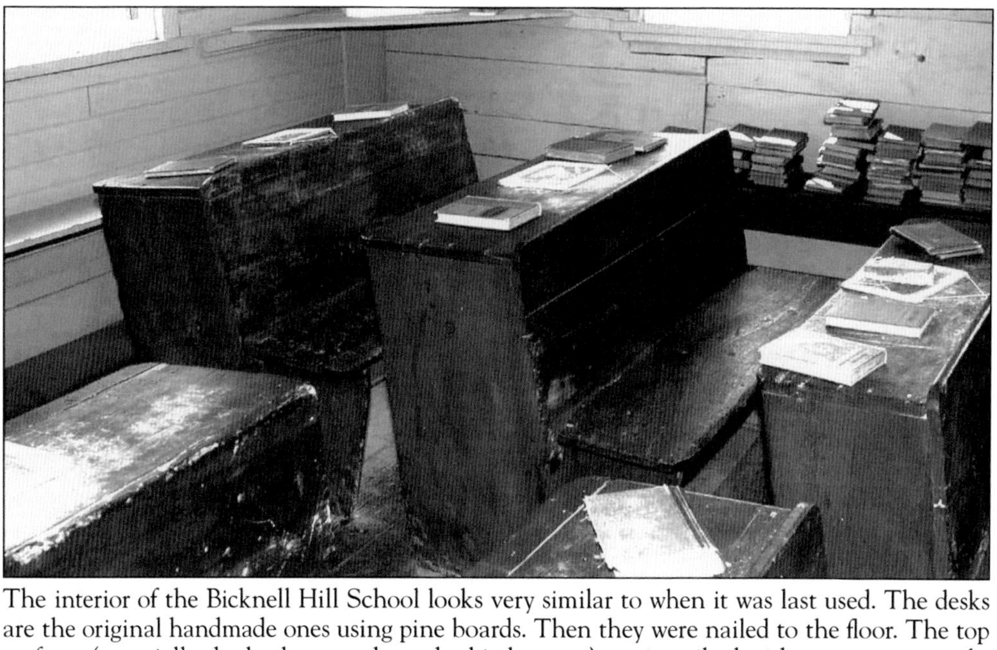

The interior of the Bicknell Hill School looks very similar to when it was last used. The desks are the original handmade ones using pine boards. Then they were nailed to the floor. The top surfaces (especially the back ones where the big boys sat) are inscribed with numerous initials. The building is also furnished with a North Tunbridge Foundry cast-iron wood-burning stove. All of the books came from Tunbridge one-room schoolhouses. (Courtesy of Mark Dixon.)

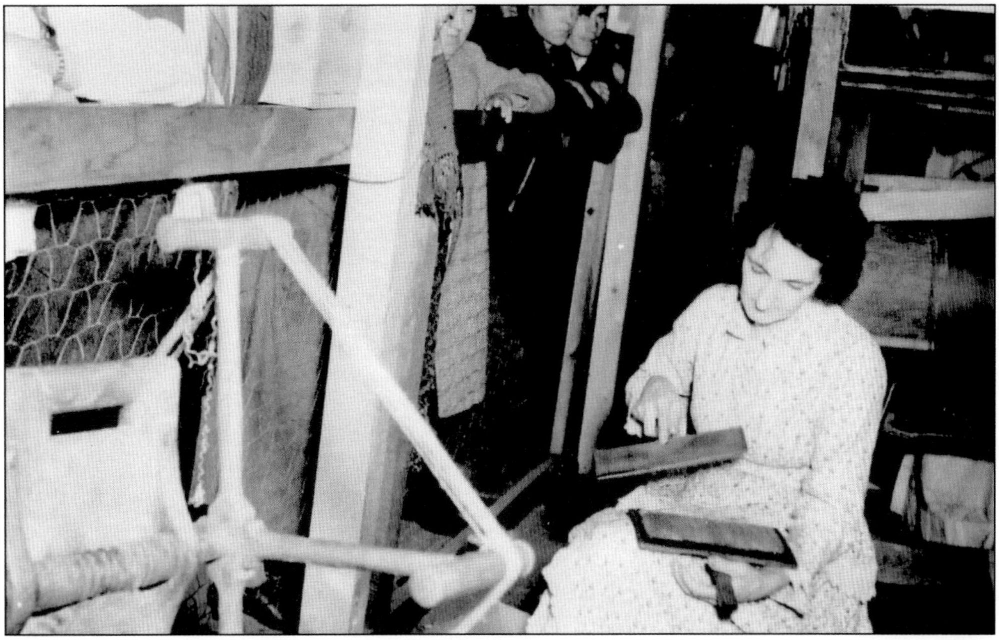

Gladys Leighton of South Royalton, watched by a good-sized audience, demonstrates carding wool. The museum has an extensive collection of spinning and weaving tools. One of the important missions of the museums' interpreters is explaining their craft.

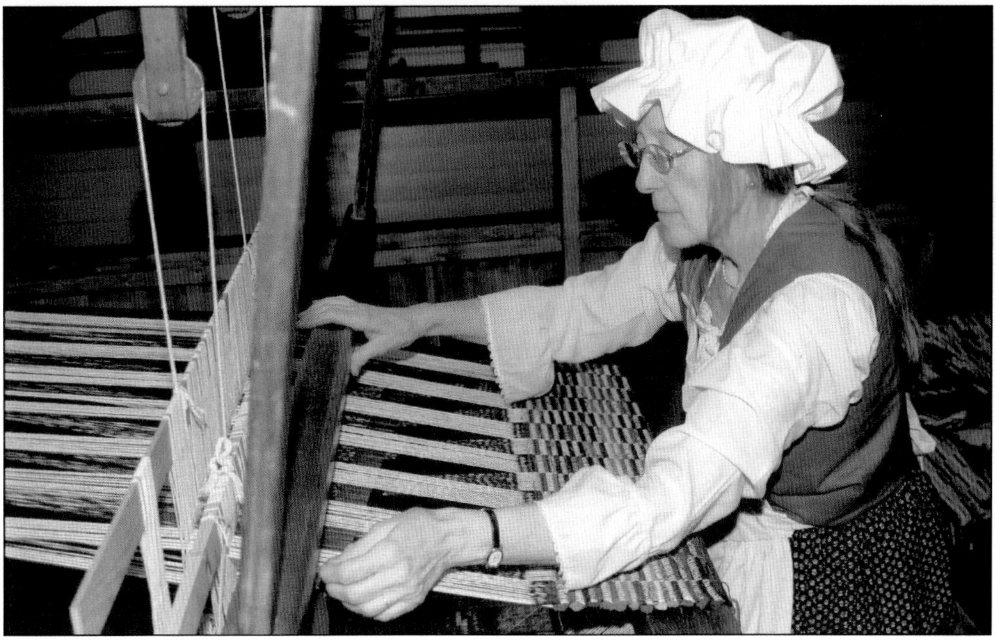

The large loom in the museum was owned by Mary (Dodge) Gould of Chelsea around 1835. Later it was the property of her daughter-in-law Emma (Roberts) Gould. About 1867, it was moved to Tunbridge. Still later it was the property of Penelope Roberts. In 1929, Edward Flint added it to the fair's collection. Over the years it has been used to weave blankets, carpets, towels, and table cloths. Jean Wolfe, pictured above, now weaves rag rugs at fair time. (Courtesy of Mark Dixon.)

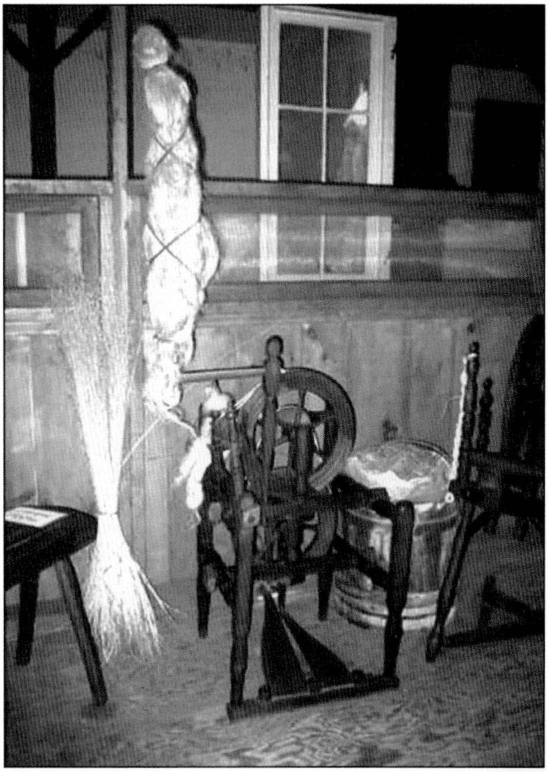

A Connecticut chair wheel was used to spin flax. The double wheels and treadles enabled the spinner to do more consistent work faster.

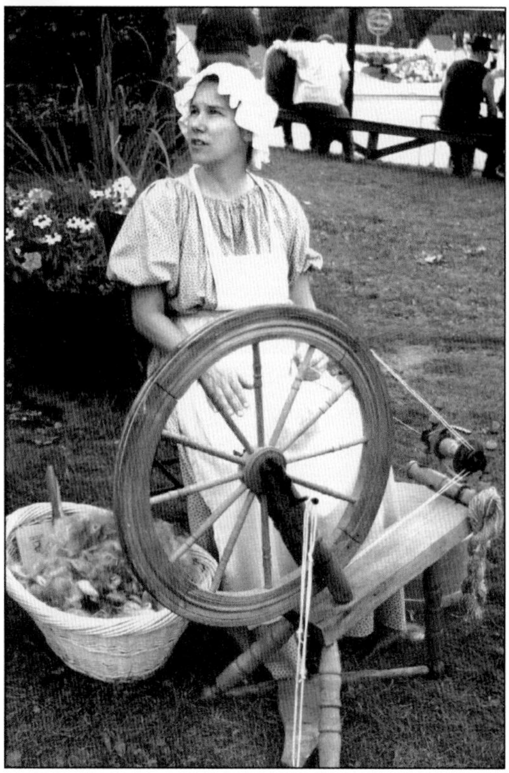

Since the weather is nice Laura Craft is demonstrating spinning sheep's wool outside the museum. Fairgoers appreciate this because they can be closer when talking with the experts. (Courtesy of Nancy Cassidy.)

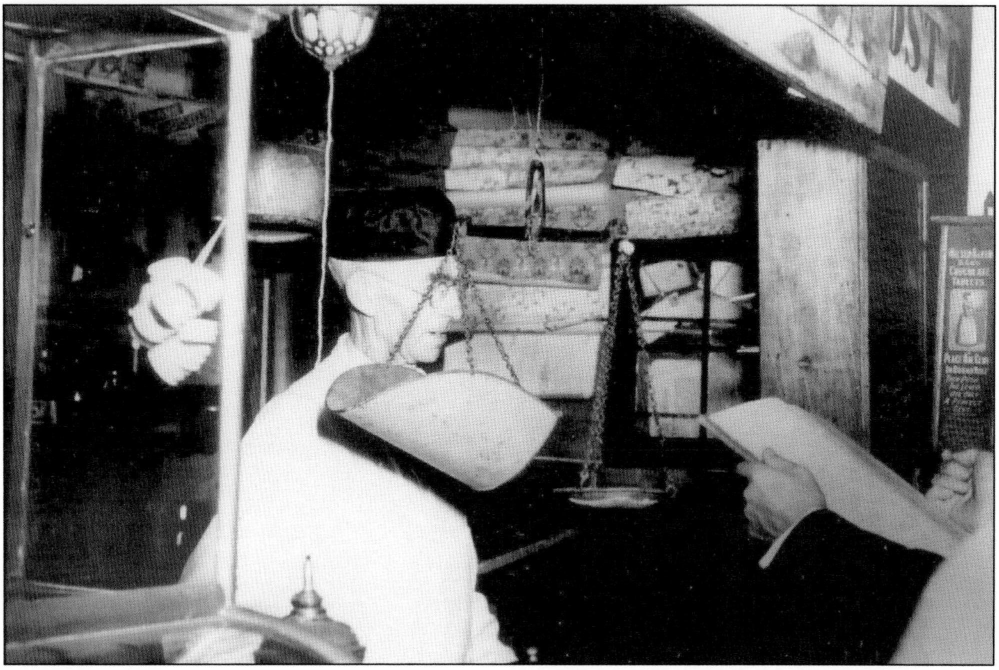

A 1940s storekeeper (notice his skull cap) demonstrated an old weighing device standing behind a chicken wire fence. The original cabin was dark, crowded and primitive; however, it was a huge success.

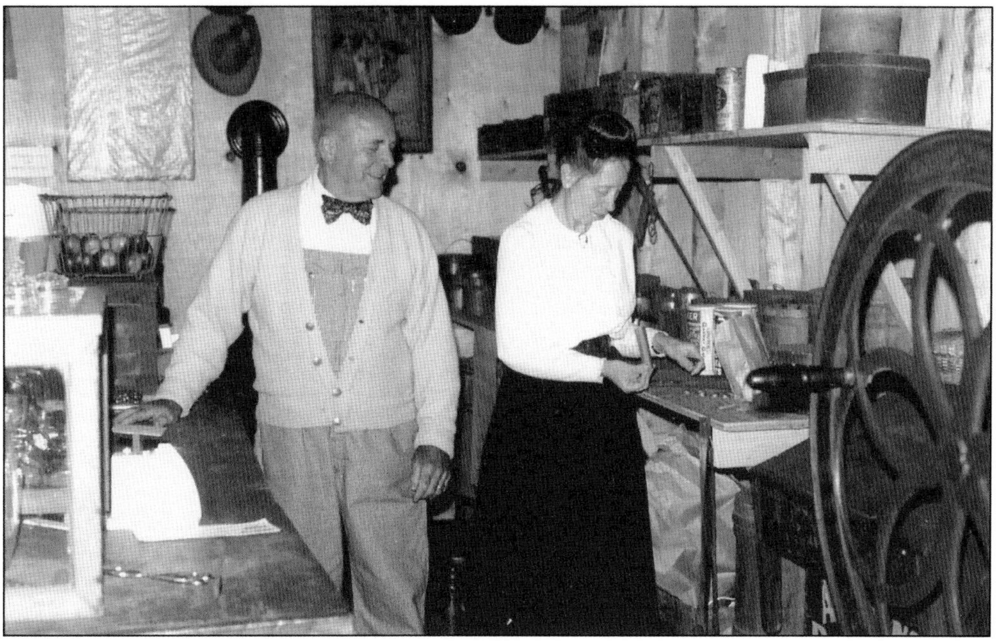

George Savidge and Priscilla Farnham, the current superintendent, are shown running the store in the mid-1980s. Previously chicken wire and plywood primarily had to be used. About this time, Priscilla Farnham began the task of redesigning the various display sections one-by-one. She guessed the whole project would take about 10 years; but, it is now 25 years and she is still at it. (Courtesy of Arlene Stockwell.)

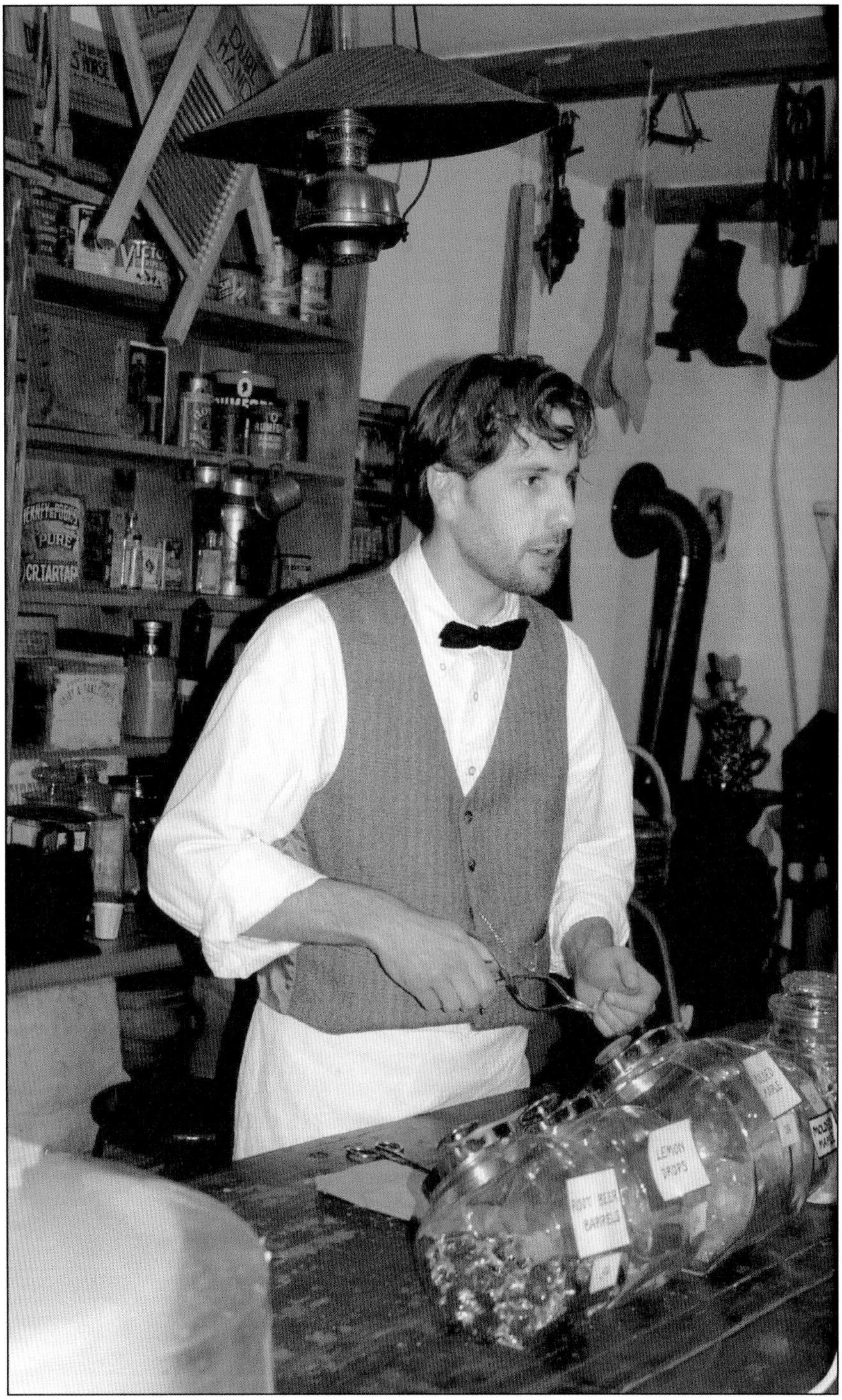

Scott Terami is about to make a candy sale. The store is usually quite busy, especially on Thursday morning of the fair when thousands of schoolchildren attend Agricultural Education Day. Hungry fairgoers can be satisfied with the following foods that would be found in a c. 1900 store: wheel cheese, Vermont crackers, pickles, apples, homemade cookies and donuts, and various candies. (Courtesy of Nancy Cassidy.)

Now the south end of the museum is a milk-to-butter display including cow stanchions, cream separators, milk cans, milk bottles, churns, and a butter worker.

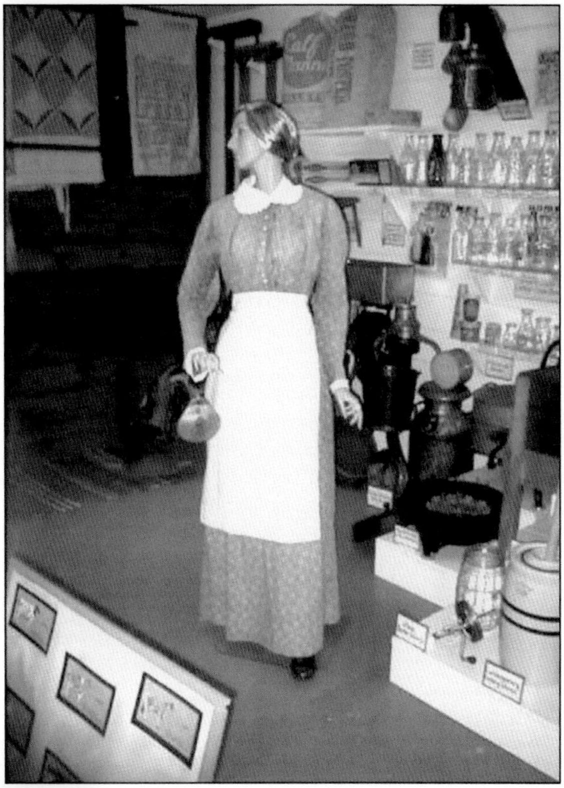

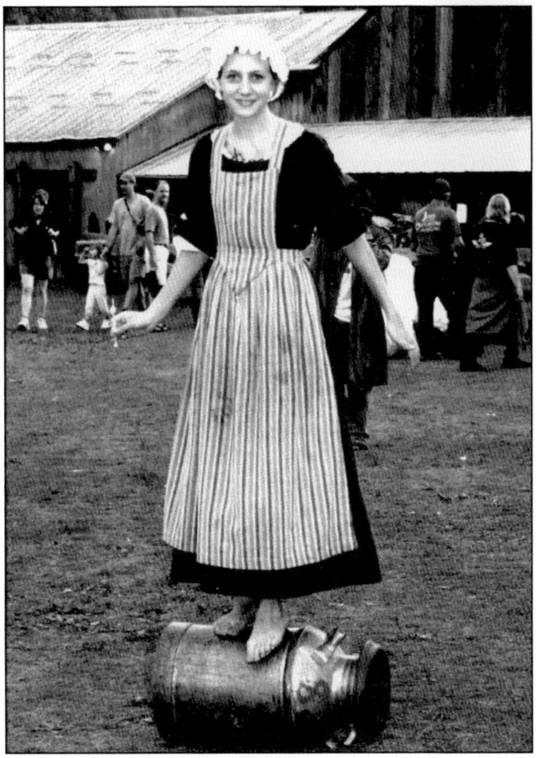

Andrea Russell started working with her mother Eve in the museum keeping room (kitchen) when she was five or six years old. She is currently studying culinary arts. During her breaks she always enjoyed milk can rolling. (Courtesy of Nancy Cassidy.)

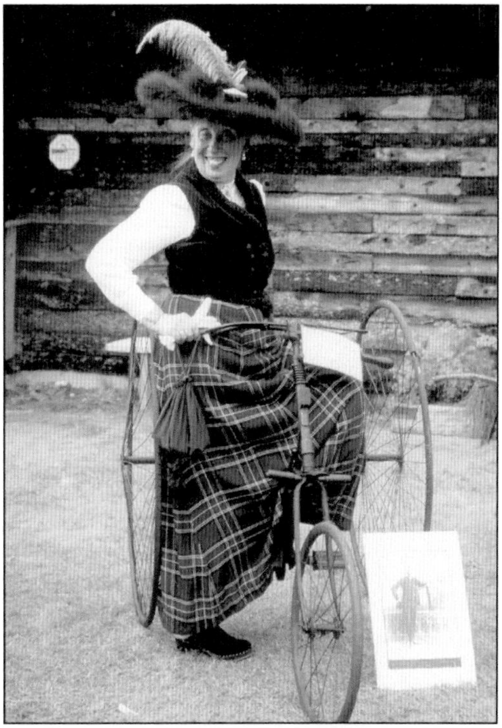

For 25 years Priscilla Farnham has headed up the Log Cabin Museum. The transformation of the displays inside has been impressive. As one can clearly see she loves hats.

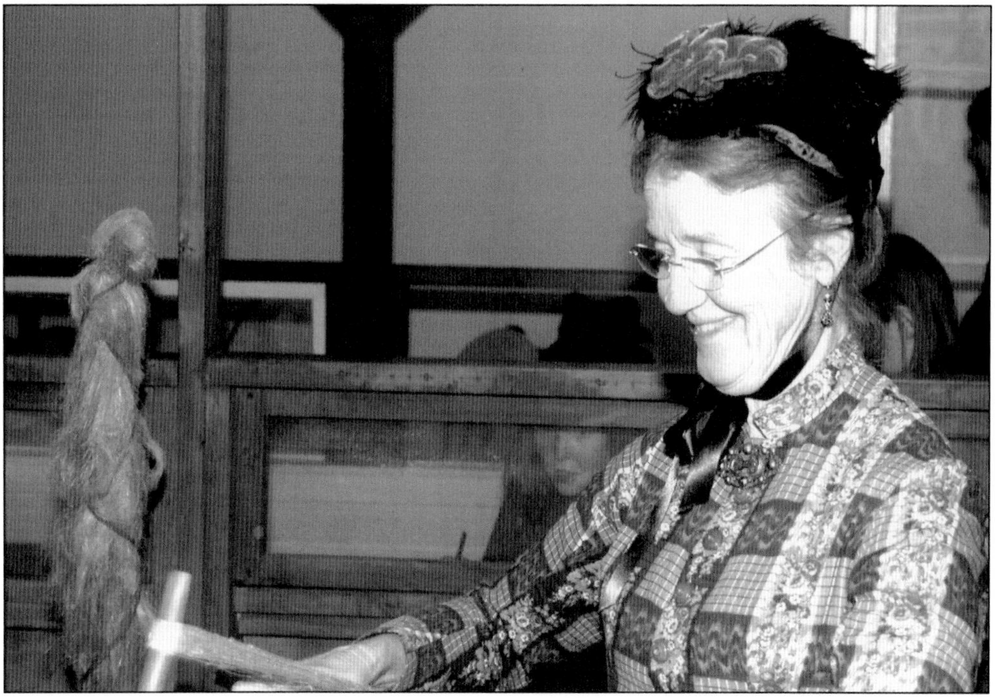

Susan Cain of Royalton (in historically accurate dress with original hat that she wears when performing with the Larkin dancers) continues the tradition of preparing flax for weaving. She is assistant superintendent of the museum and schoolhouse, as well as an expert on the history of domestic arts. (Courtesy of Mark Dixon.)

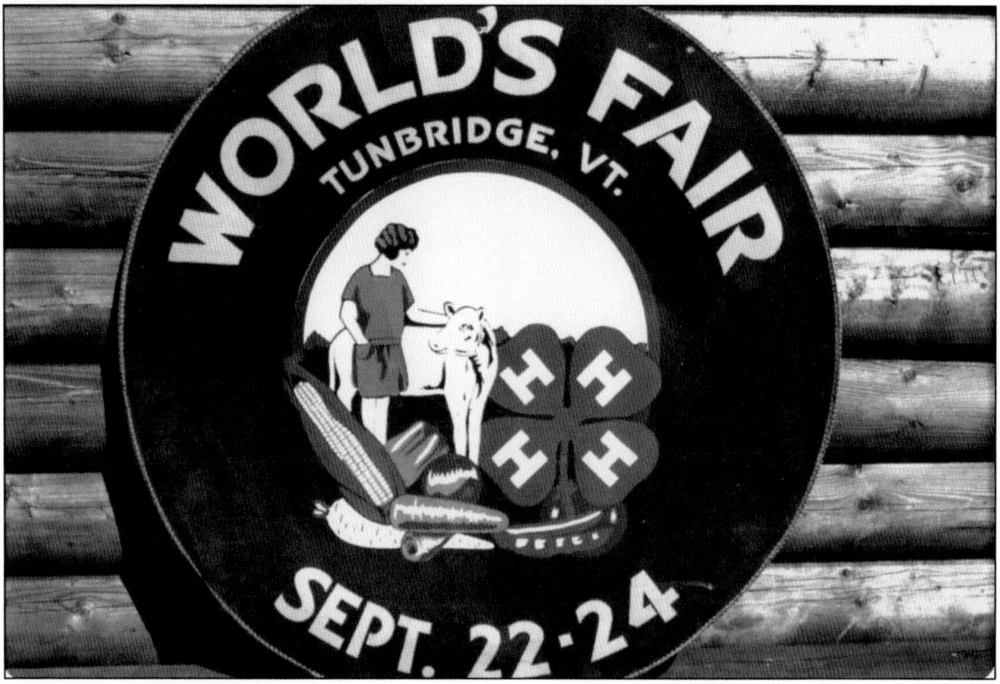

In the late 1920s, the fair decided on a novel method of advertising using a spare-tire cover, the first attempt at publicizing the fair using automobiles. Another style is displayed in the museum. Such items are now pricey antiques sought by collectors.

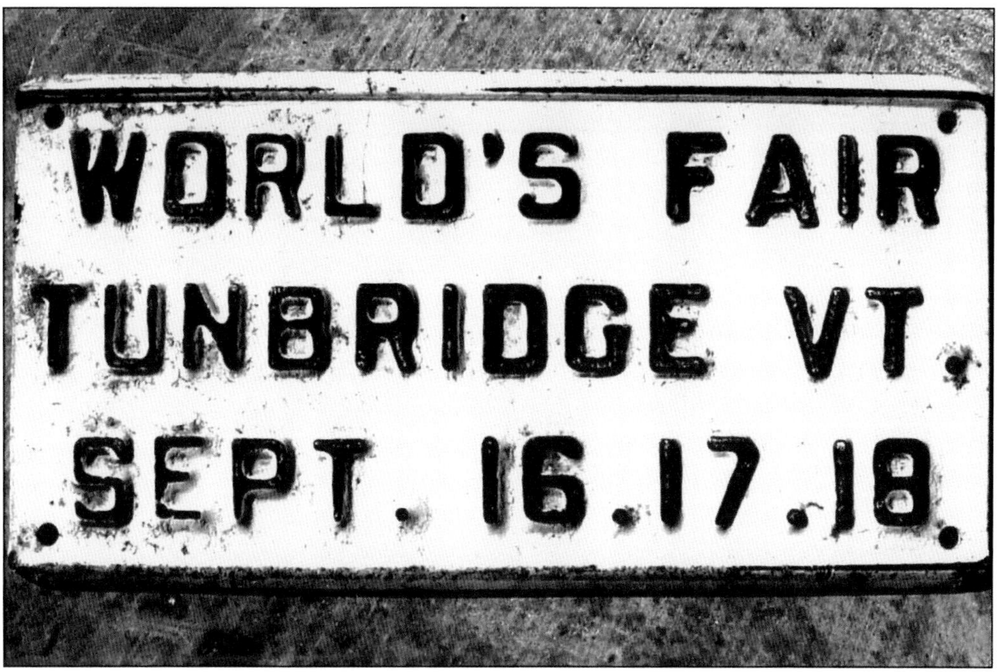

Yearly the fair advertised with metal plates for 10 to 12 years. The last was in 1941. Today they are rare collectables.

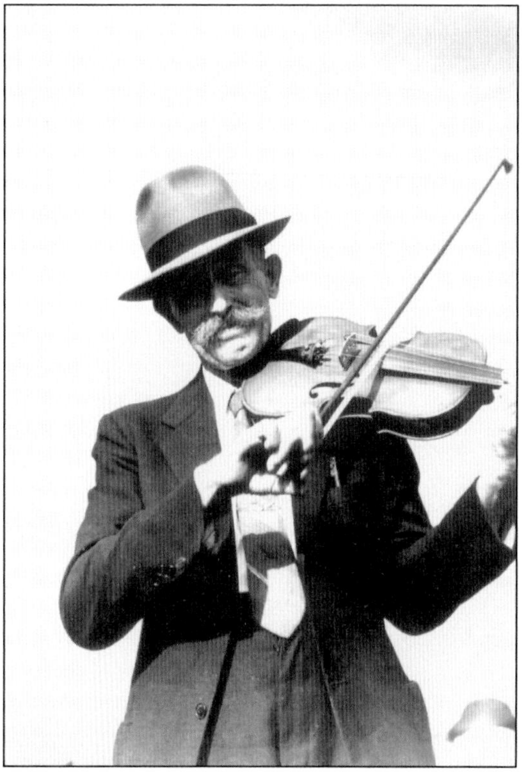

Ed Larkin founded the contra dance group in about 1934. They have performed at almost every Tunbridge Fair since then. Larkin was a tough taskmaster who would stop the troupe in mid-dance if someone made a mistake. He was a farmer; however, his real loves were fiddling and calling dances.

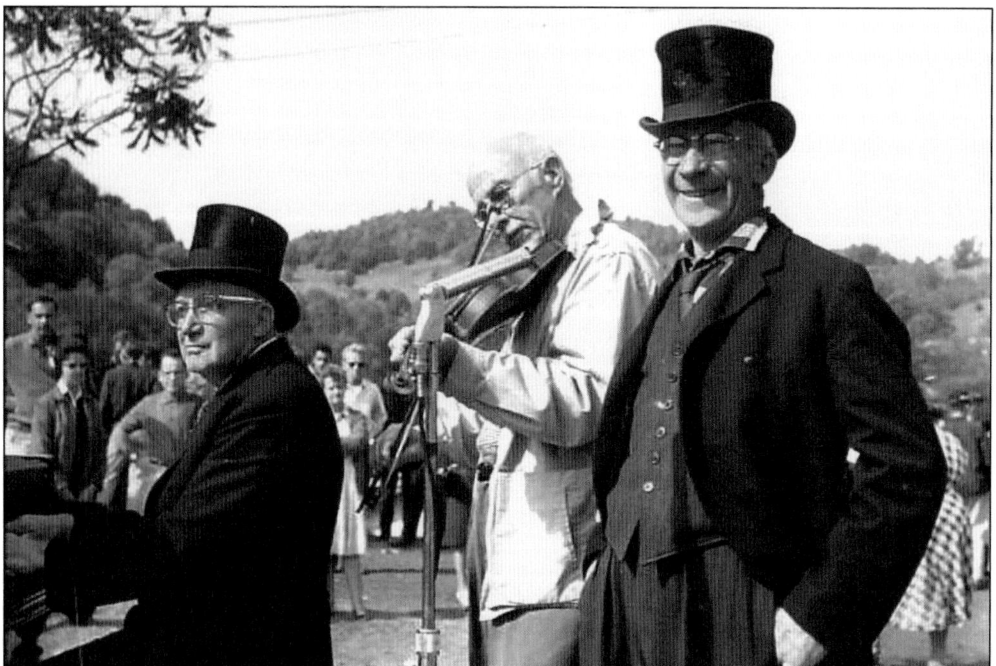

Walter Sawyer, left, and Max Hayward provided music for the contra dancers. Harley Ladd was the caller. One of the factors that made Larkin's group unique was their costumes. Max Hayward alone always refused to wear a costume.

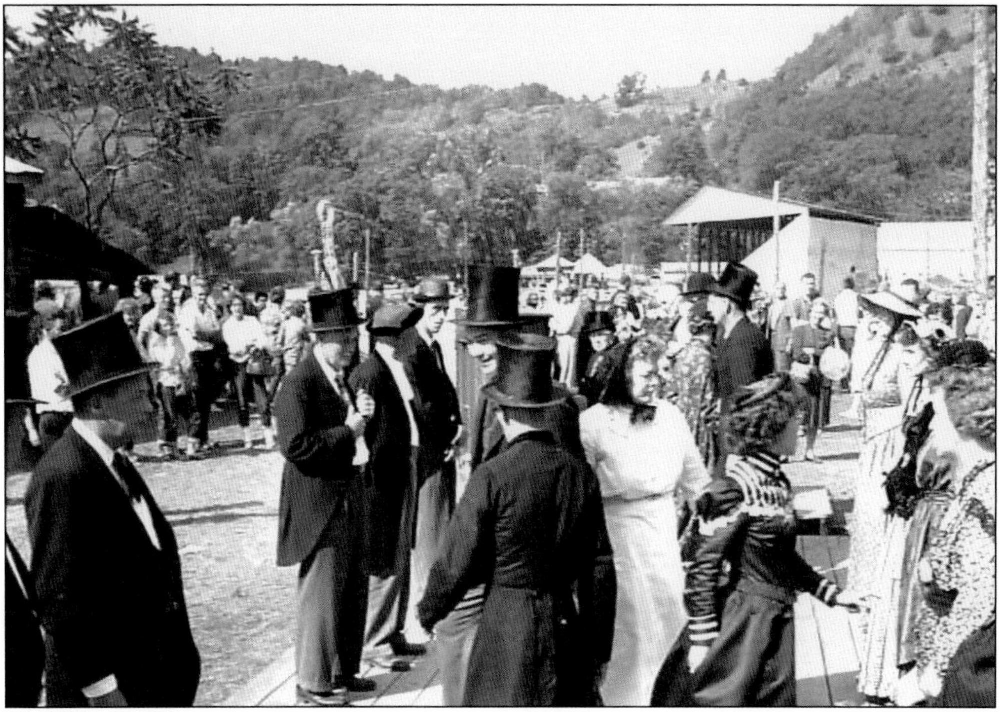

For years the contra dancers performed on the green in front of the museum. If it rained this meant that they could not go on. Later the group danced there under canvas. Today the troupe headlines the acts in the fair's entertainment tent.

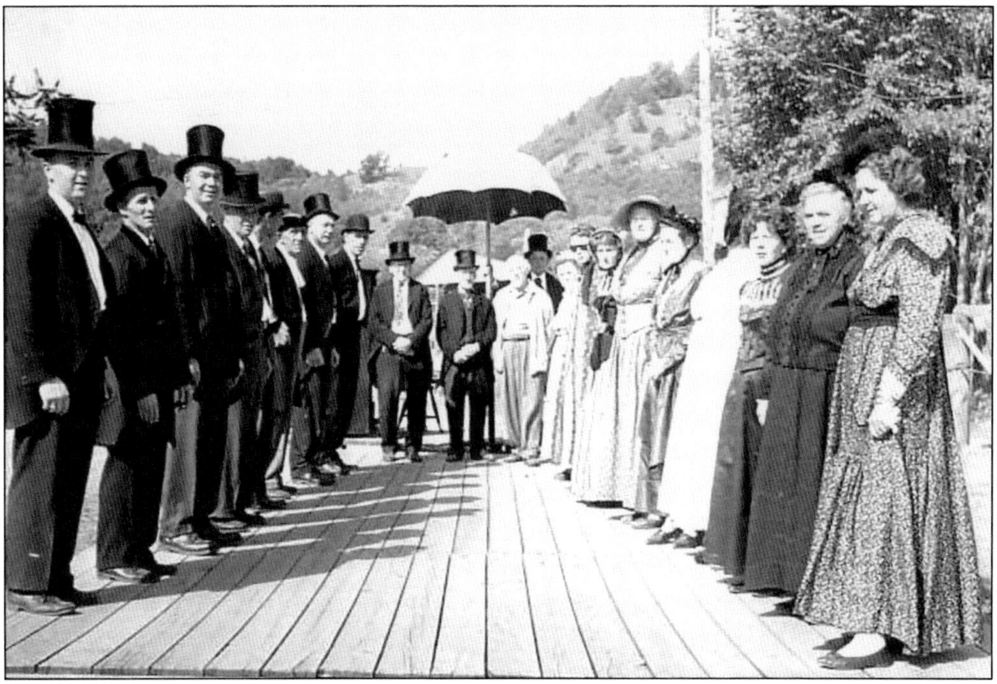

The Ed Larkin Contra Dancers pose for an official portrait during a 1960s fair. They all lived in the Tunbridge area and had been together for years.

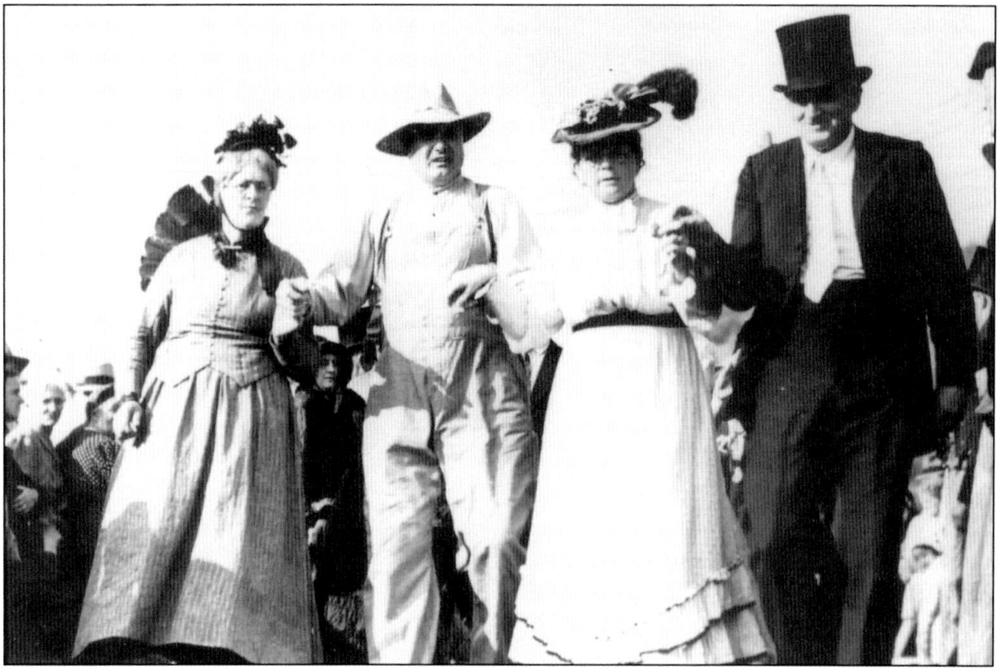

The most interesting contra dancers of years past were Alice and Forrest Hood, left, and Charles and Mary Durkee. The Hoods dressed as an old farm couple. At least once during each performance Alice managed to "drop her drawers." She then loudly berated her husband. Charles and Mary were known as two of the smoothest dancers on the floor.

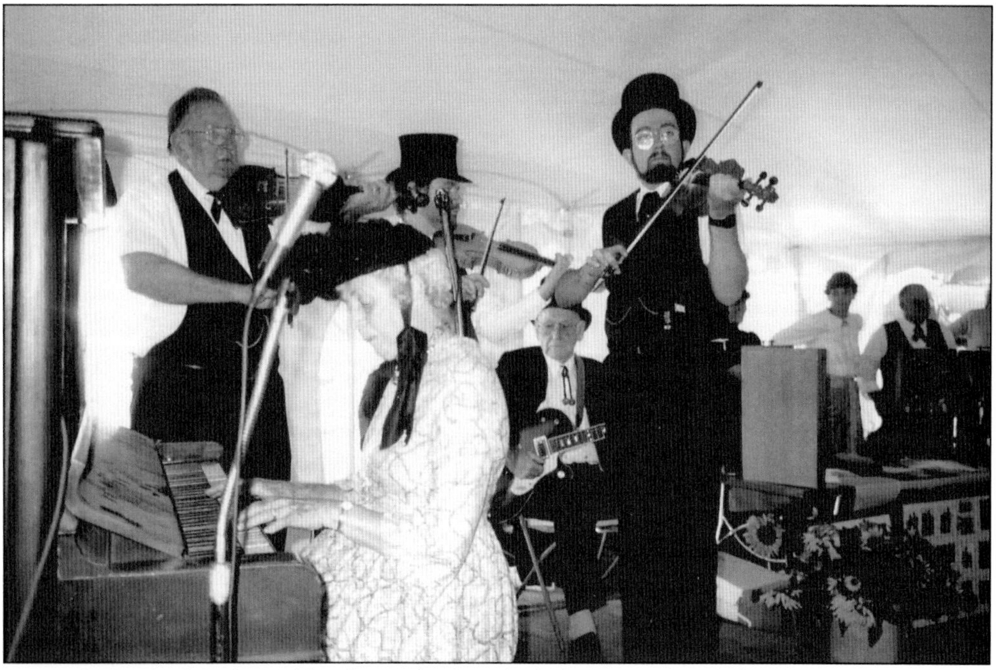

Gertrude (Larkin) Roberts (daughter of Ed Larkin) for years played chords at the piano. Bill Roberts is shown playing the guitar. The fiddlers are Wayne Doyle, left, Mike Jennings, center, and Adam Boyce. (Courtesy of Byron Angell.)

Three

Livestock Shows

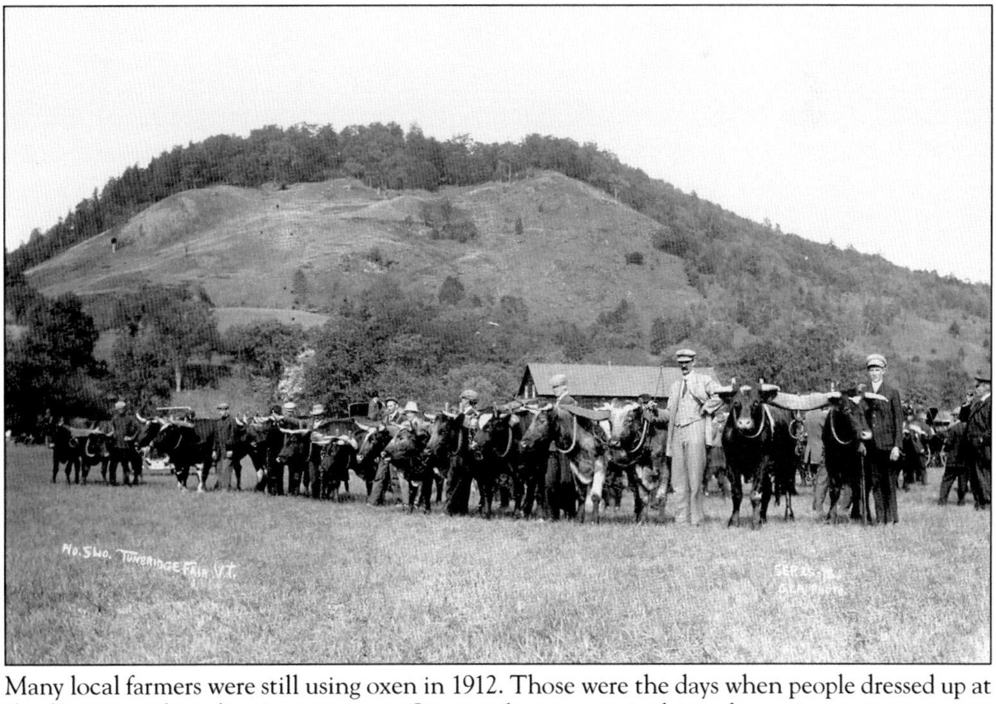

Many local farmers were still using oxen in 1912. Those were the days when people dressed up at the fair, even when showing ox teams. One gentleman is attired in a three-piece suit.

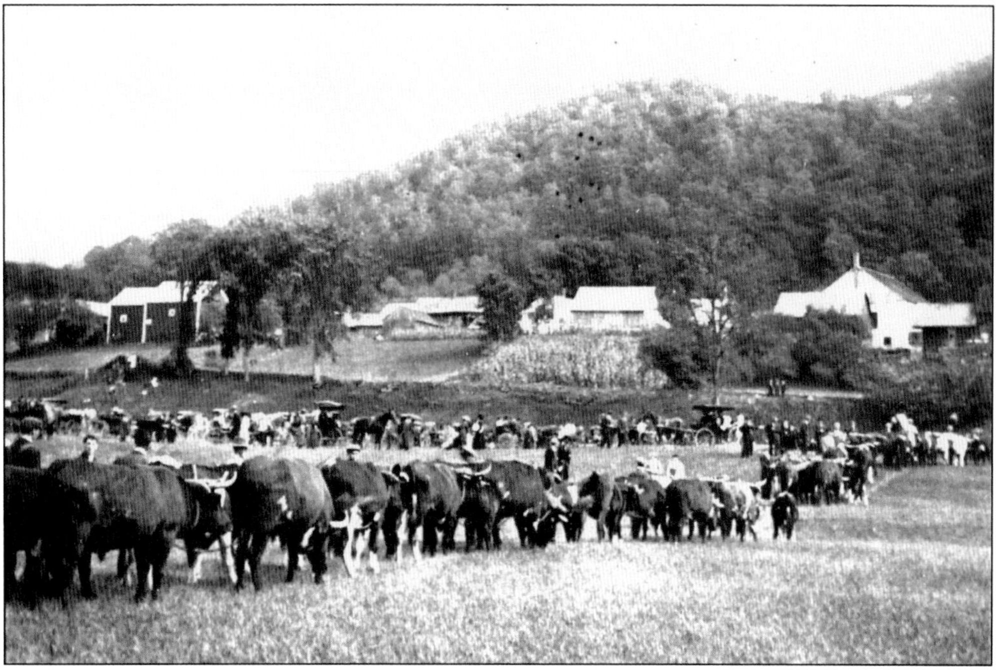

Most of the oxen assembled in 1912 in the long line were from the First Branch Valley farms. In the early years in this country most oxen were Durhams.

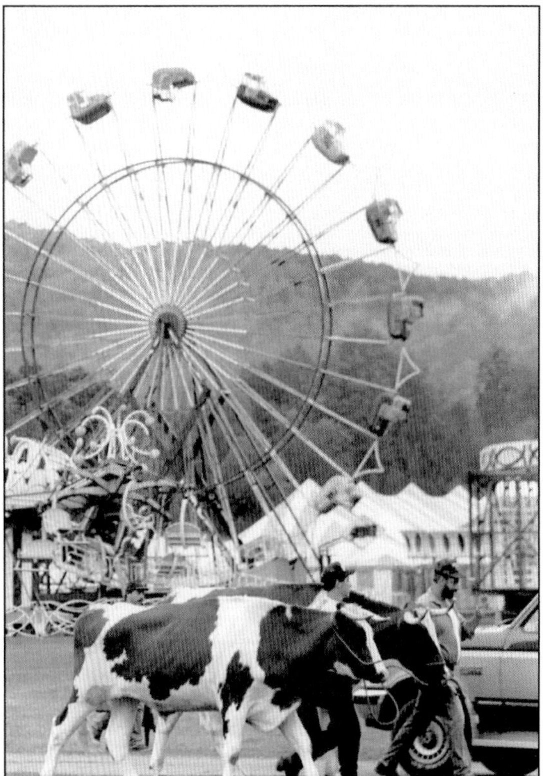

The old and the new were represented at a recent Tunbridge Fair. Brian Hayward, right, of Chelsea exercises his ox team in the early morning along the midway. Teamsters enjoy showing off their animals. (Photograph by Bob Eddy.)

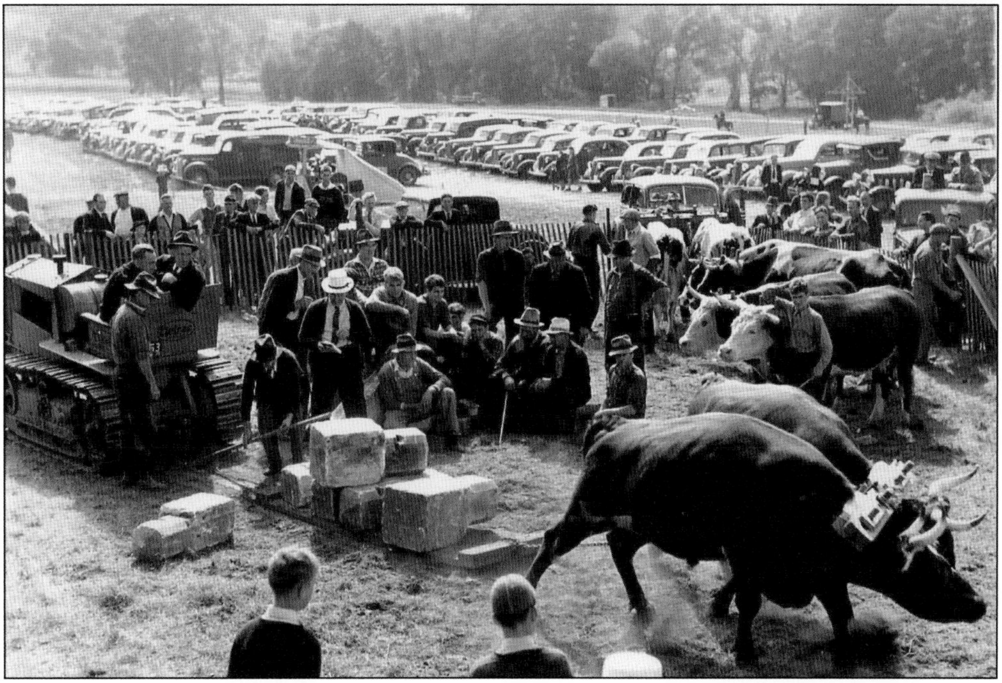

Oxen strain to pull the load of cement blocks in this pre–World War II photograph. The automobiles in the background help date the scene. The measurer waits behind the blocks in order to record the distance.

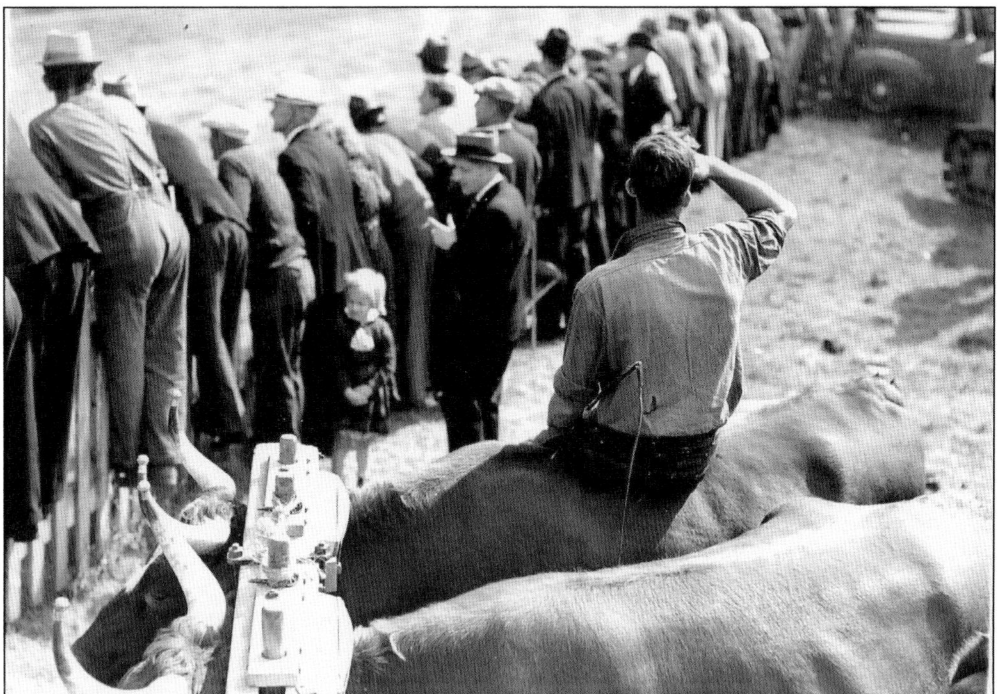

Back in the days when pulling events were held outside, it was difficult to find a good vantage place from where to view the event. So why not sit on the back of an ox?

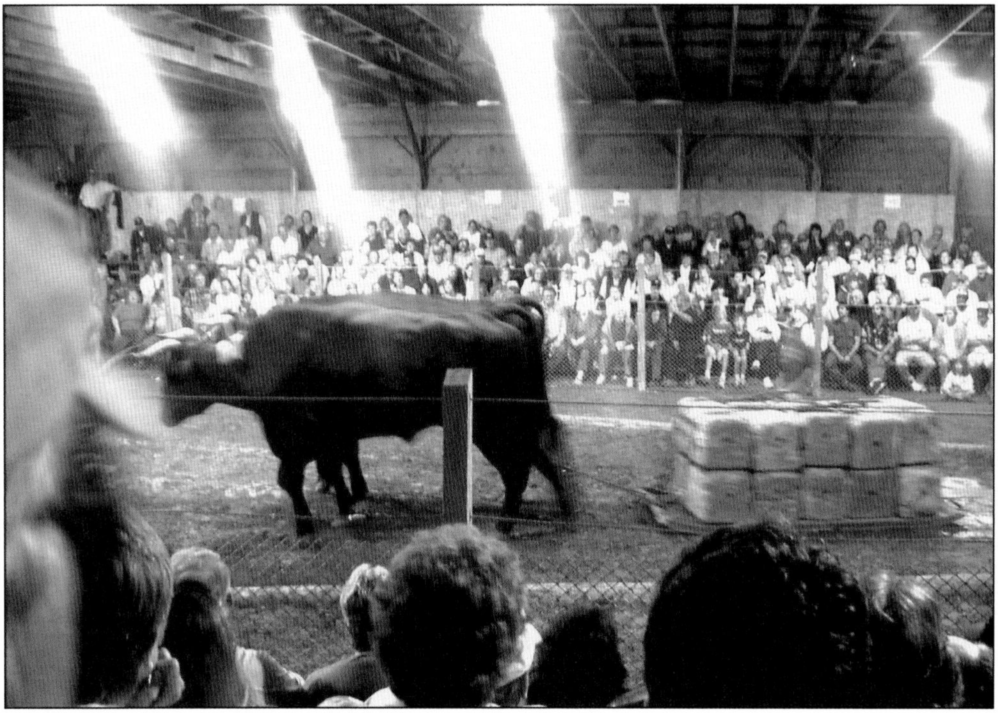

When pulling events were held out-of-doors sometimes they were rained out. Never again will this happen! The Memorial Arena was constructed in 1989. Now every seat in the building is usually occupied. (Courtesy of Candi Barr.)

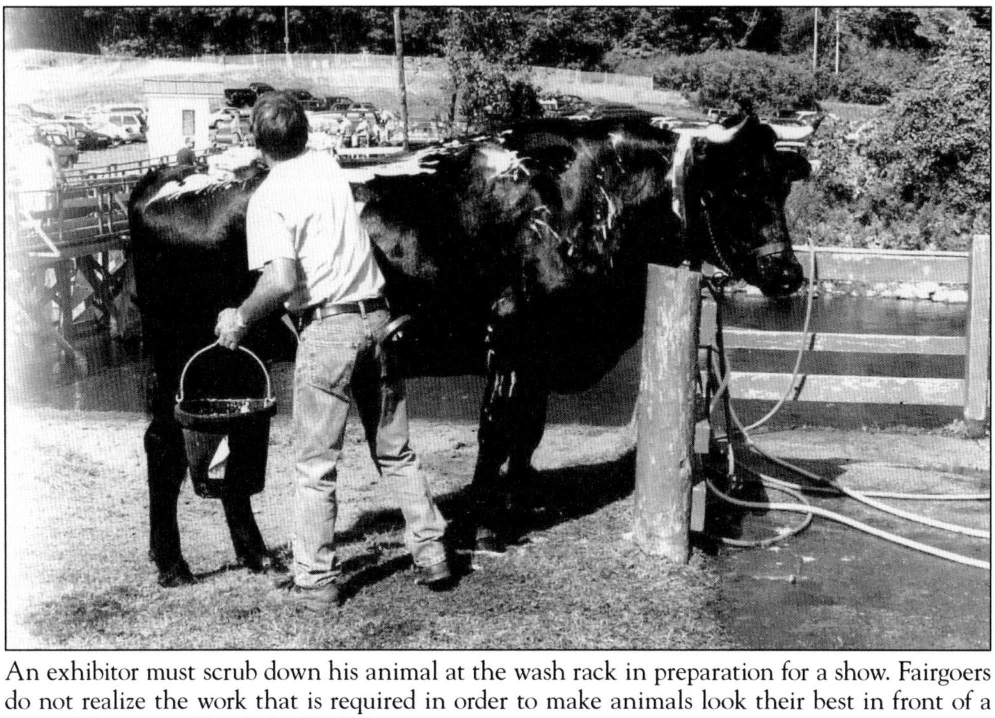

An exhibitor must scrub down his animal at the wash rack in preparation for a show. Fairgoers do not realize the work that is required in order to make animals look their best in front of a judge. (Courtesy of Nicholas Wolff.)

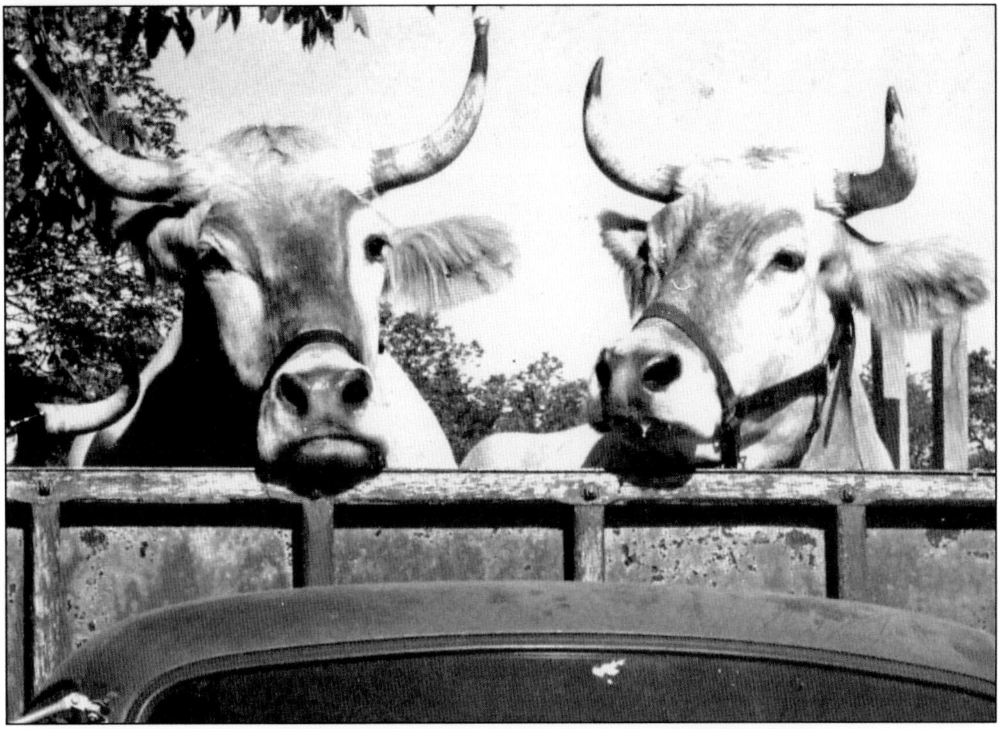

The 1963 fair was over and these oxen were loaded and ready to head for home.

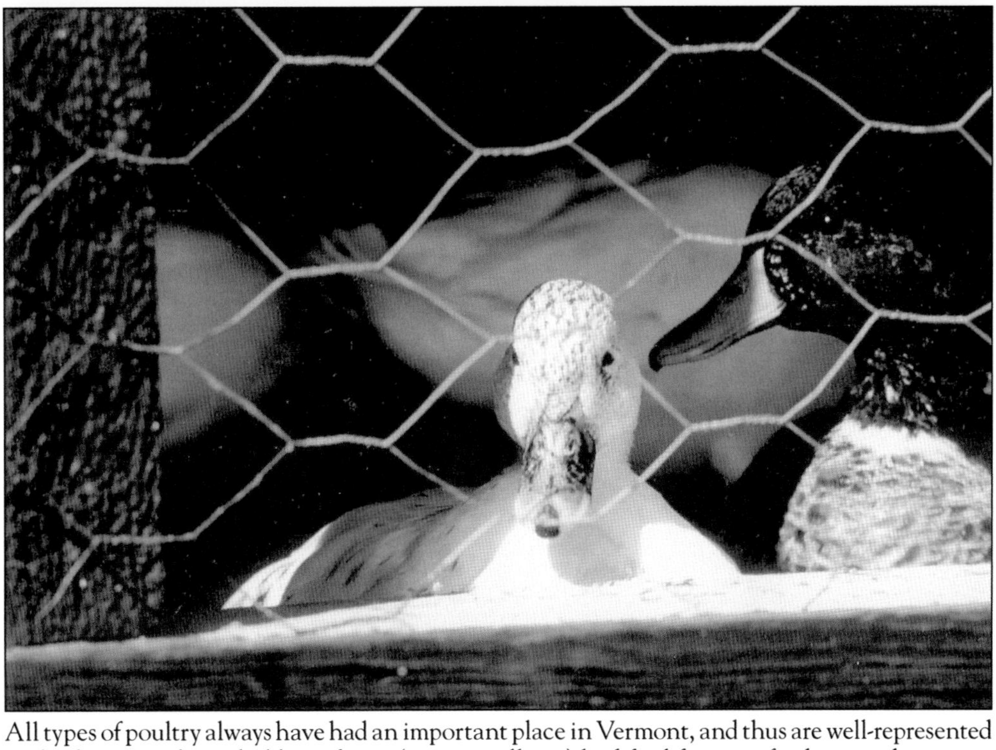

All types of poultry always have had an important place in Vermont, and thus are well-represented at the fair. Most households on farms (even in villages) had fowl for eggs, feathers, and meat.

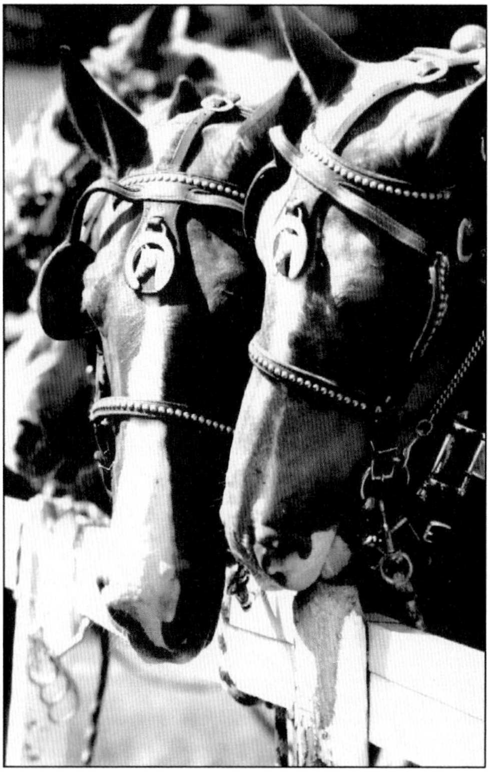

A beautiful team of pulling horses in full harness awaits its turn outside the Memorial Arena. (Courtesy of Nancy Cassidy.)

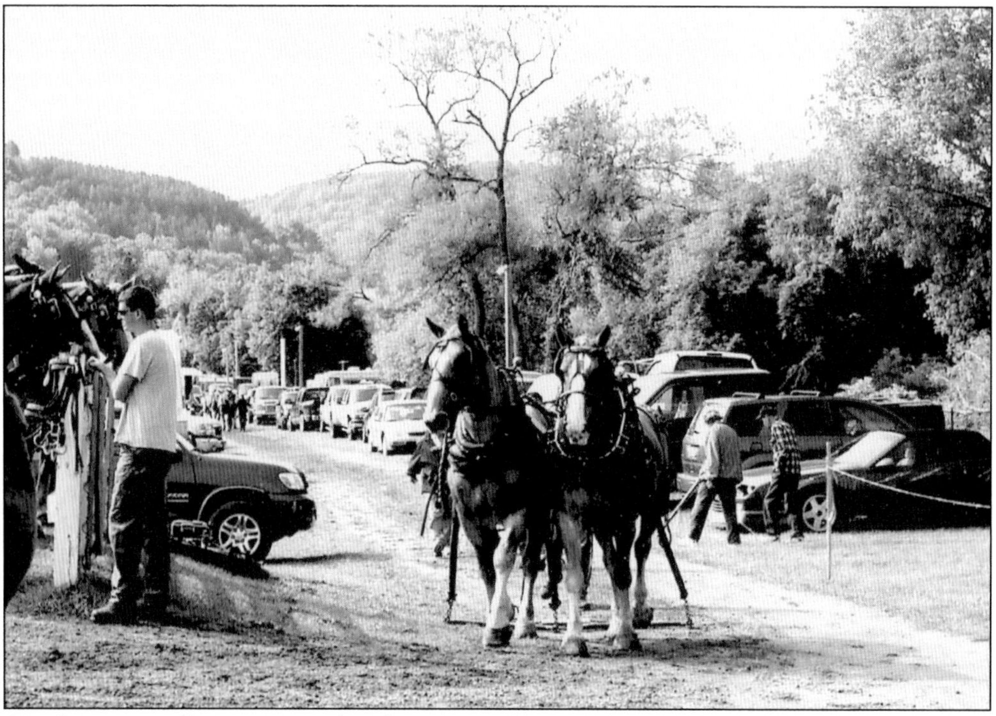

A pulling team of horses approaches the arena. Pulling events (pony, horse, and oxen) have been a popular tradition at the fair for most of its existence.

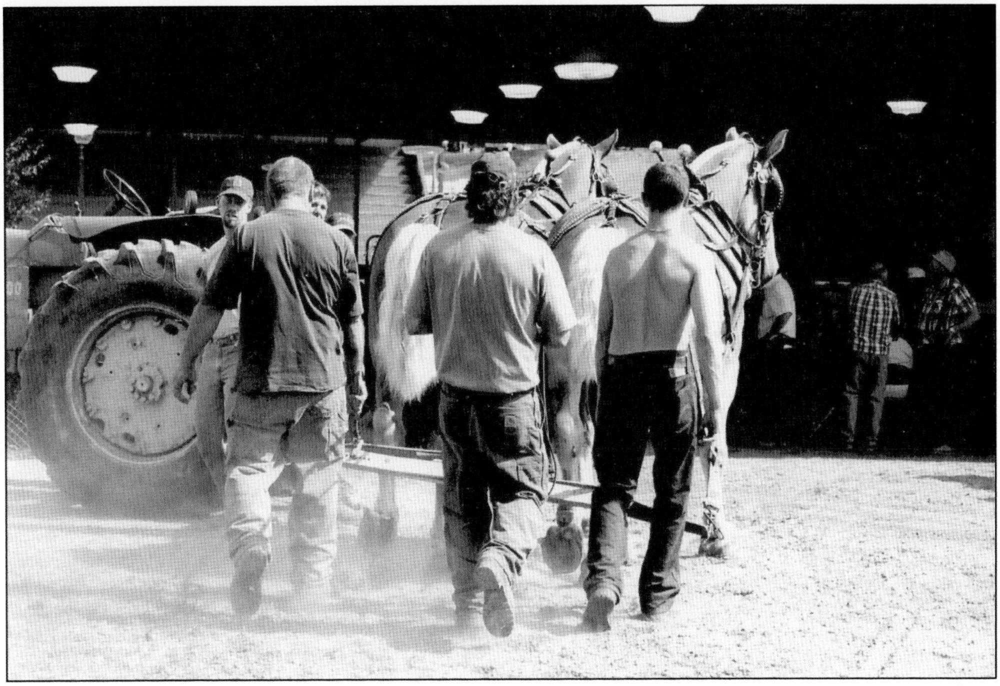

A teamster, center, and his two assistants with their team are about to enter the arena for their pulling event. Teams from all over the North Country come to this event.

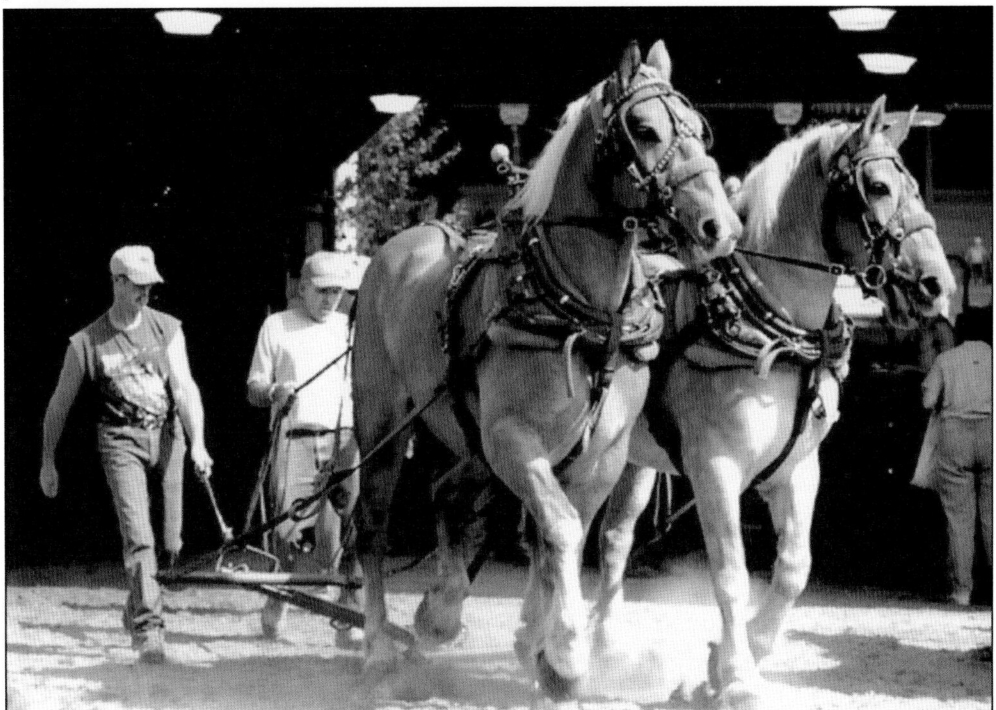

A handsome team emerges from the arena still "psyched up" and steaming from the competition. They are very large animals with a high degree of intelligence. (Courtesy of Nancy Cassidy.)

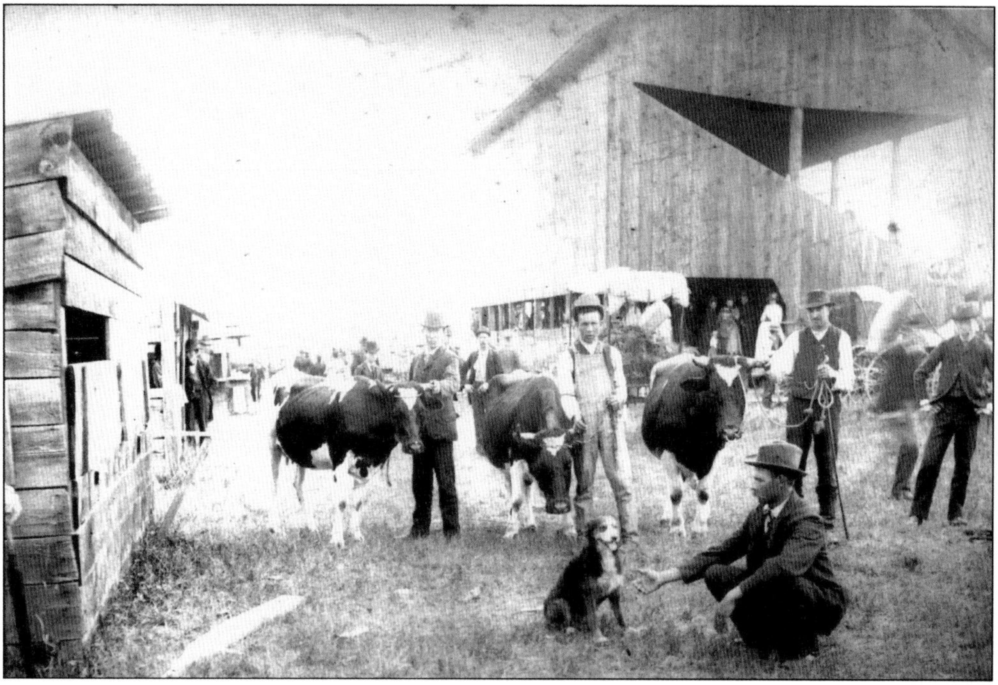

Since the cattle sheds were near the river around 1920, this photograph was probably staged at the north end of the original grandstand. It would be interesting to know the role of the dog.

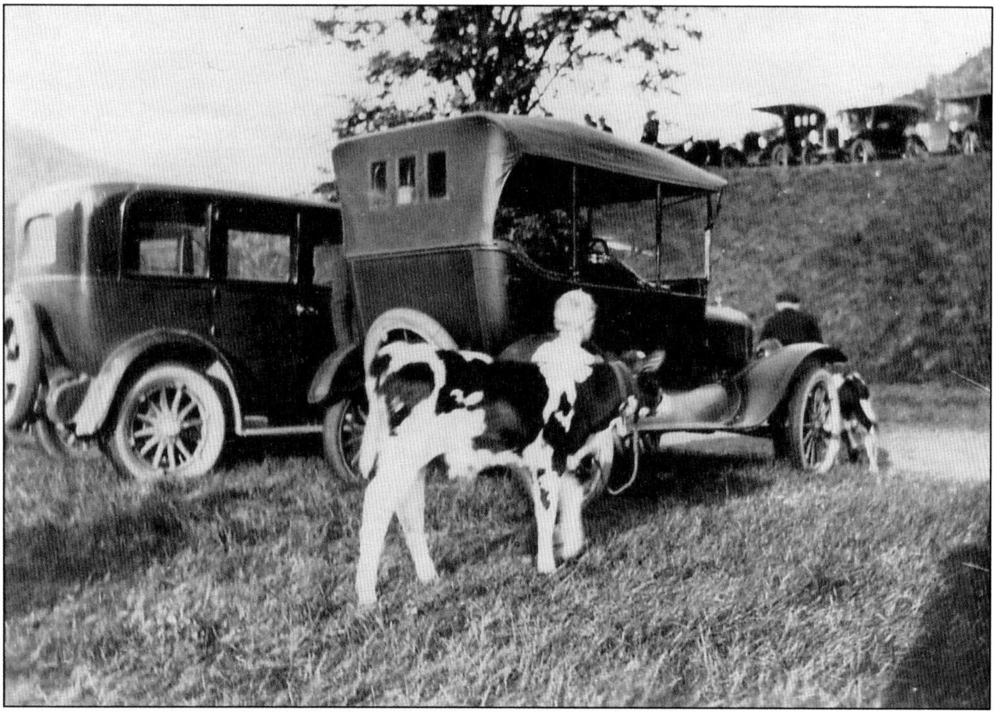

The Howe family has had a long and close relationship with the fair as well as with Holstein cattle. Six-year-old Mark leads a calf during the 1925 fair. He was the son of James K. and Marion Howe. Years later Mark became superintendent of cattle.

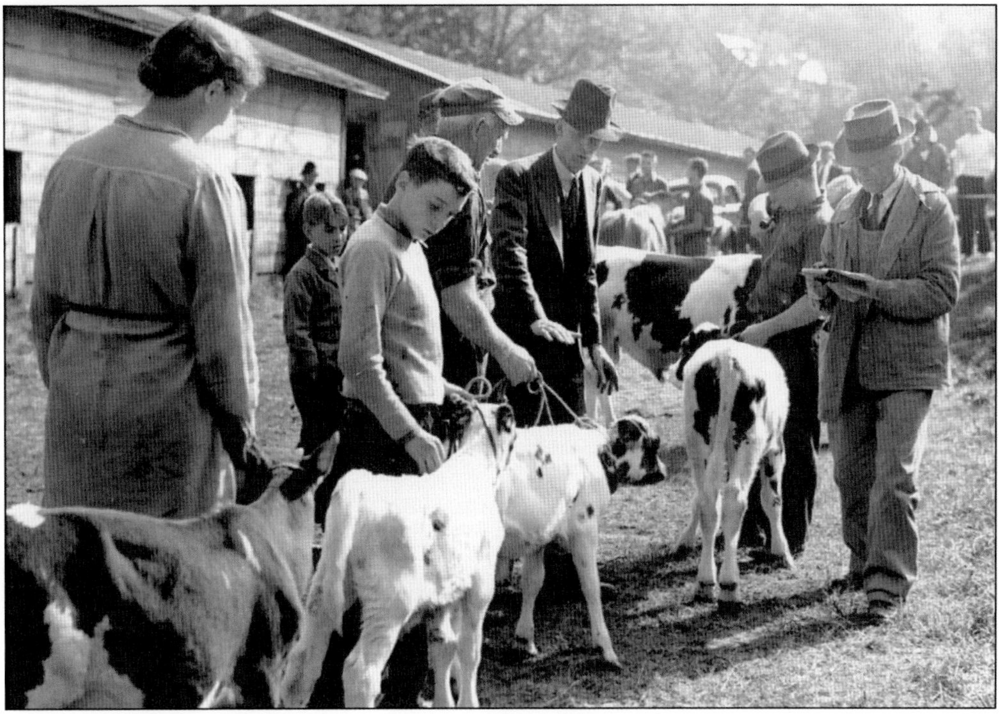

Superintendent James K. Howe is reviewing his list of show calves during a 1940s fair.

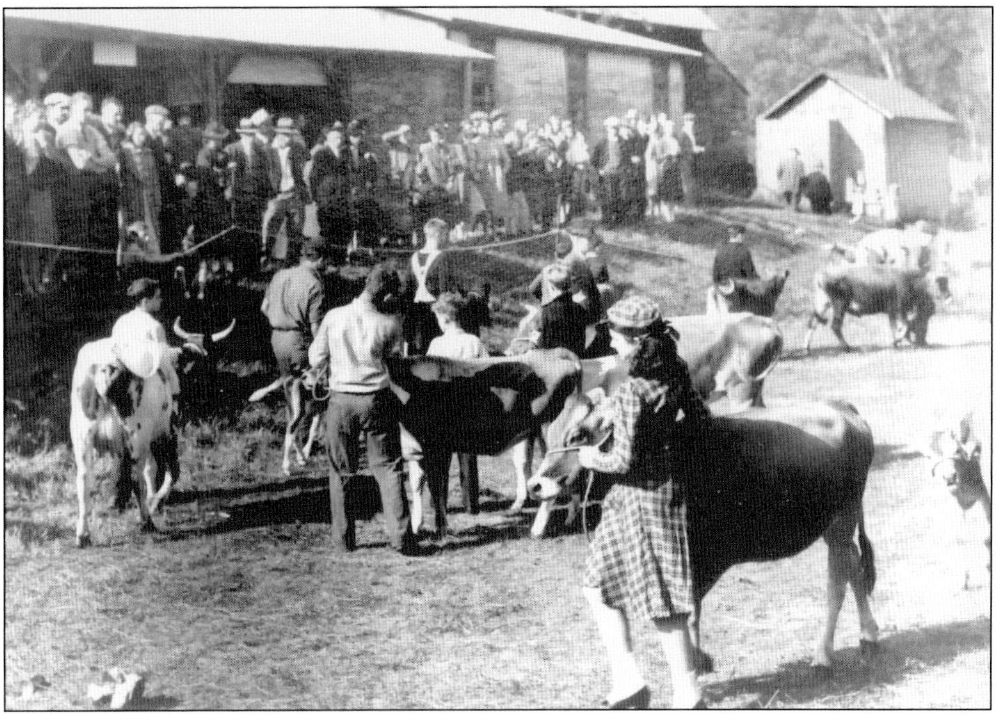

In the 1940s, cattle judging events took place between the north side of Floral Hall and their barns. A group of 4-H club young people lead their cattle before the judge. Notice the fashionably clad lass in the foreground.

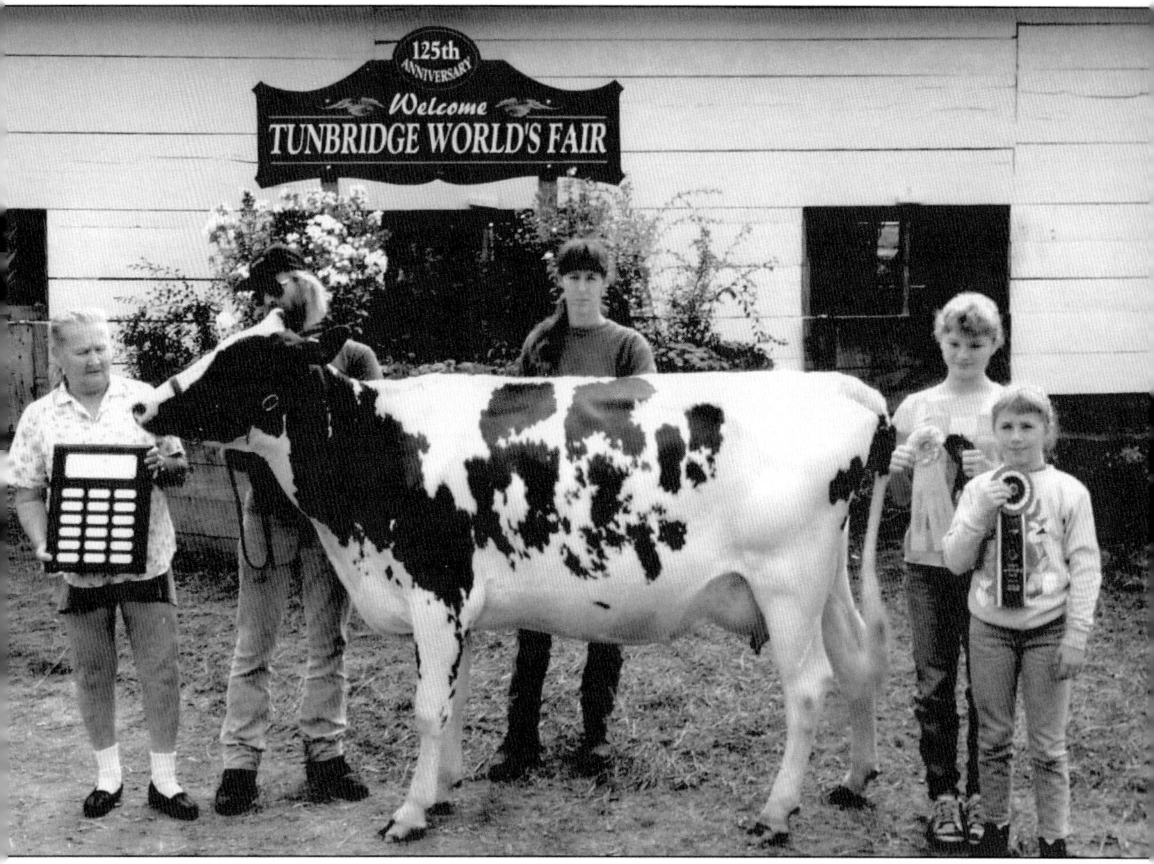

Anne Burke, left, and her family pose with their Ayrshire cow, which was named the Supreme Dairy Cow of 1996. The award is in memory of James K. Howe, a longtime dairy superintendent. Anne Burke has been a fair exhibitor for years. Harvest Hill, the family farm, received the same award both in 1996 and 1997.

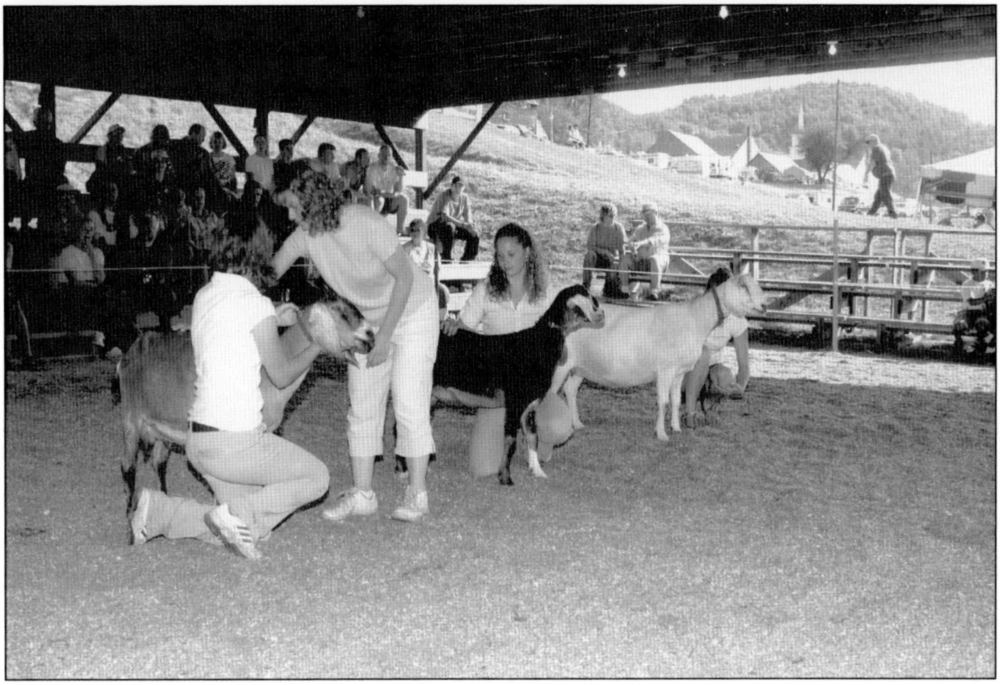

All livestock competitions—dairy cattle, oxen, sheep, and goats—are held in the judging arena. These young ladies are presenting their animals in the open youth goat show.

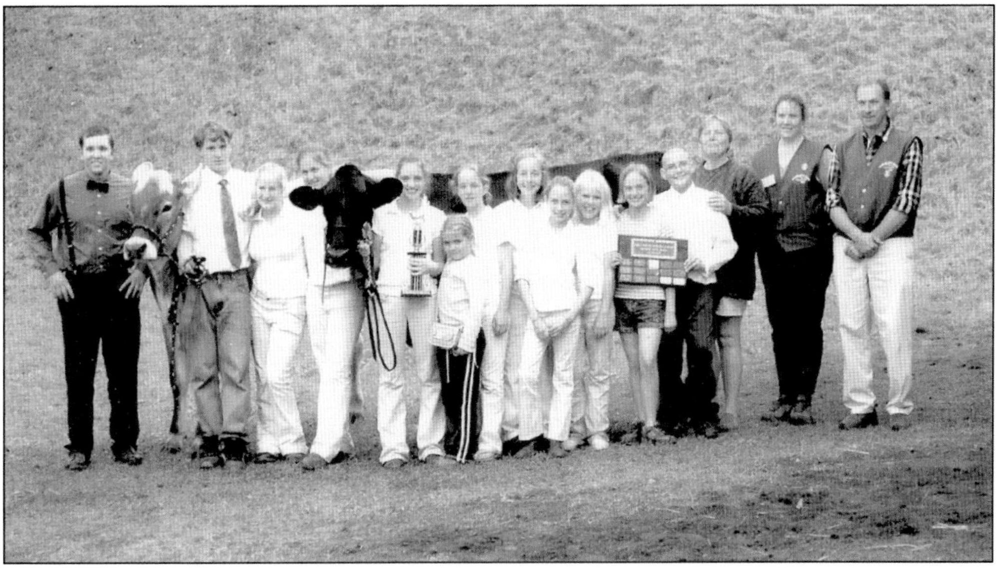

Members of the New Beginnings 4-H Club show off their trophy. Miriam Pease Kill, third from right, their proud leader, poses with Jena (Trombly) Young and Rob Howe, superintendent of cattle.

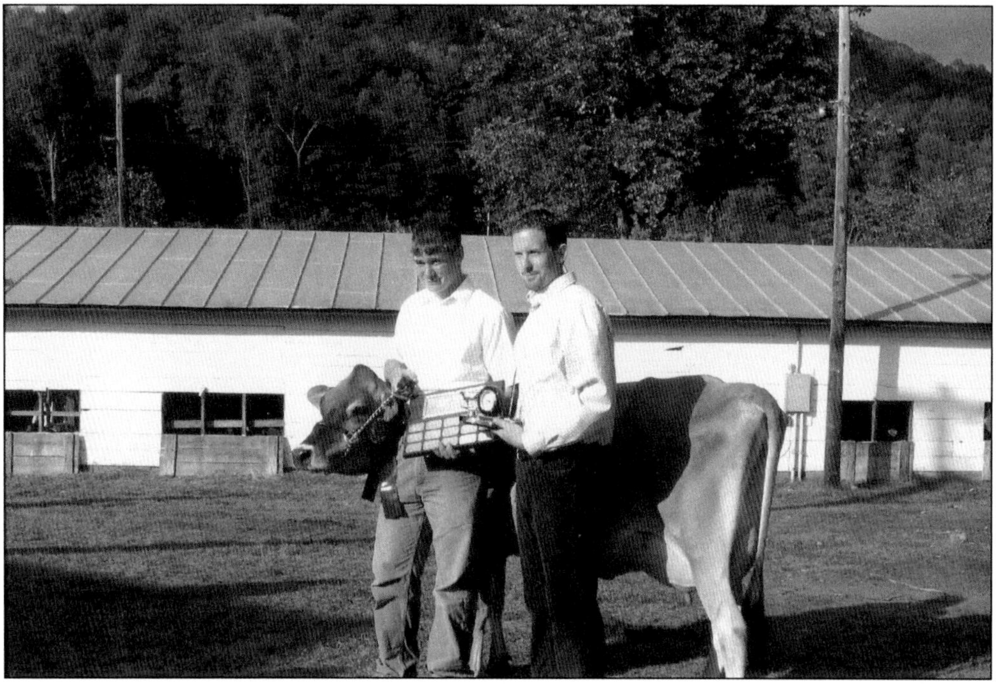

Joseph Angell, left, a Randolph resident, twice won the junior division's Walt Record Supreme Dairy Animal Award. In an era when dairying is in decline, it is heartening to see a young man proudly showing his beautiful Jersey. (Courtesy of Byron Angell.)

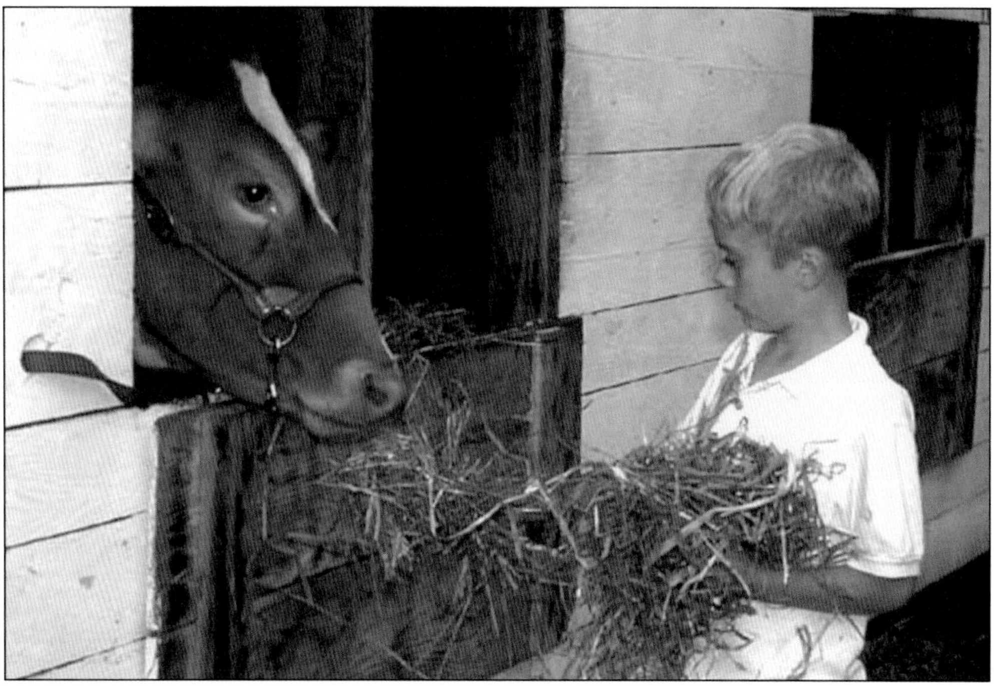

Lucien Pease takes a bit of quiet time with one of the Pease Family Farm's cows. He is dressed in his white uniform, indicating that he recently had been leading a dairy animal in the judging ring. (Courtesy of Nancy Cassidy.)

Fair photographers love to catch cattle drinking in the First Branch of the White River, which flows along the north side of the barns. This lovely Jersey is looking back at spectators on the footbridge. (Courtesy of Nancy Cassidy.)

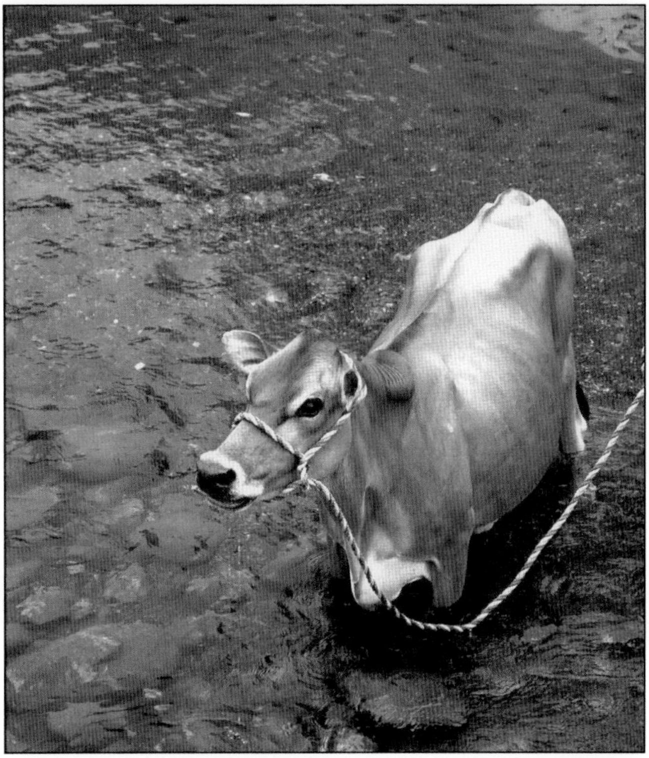

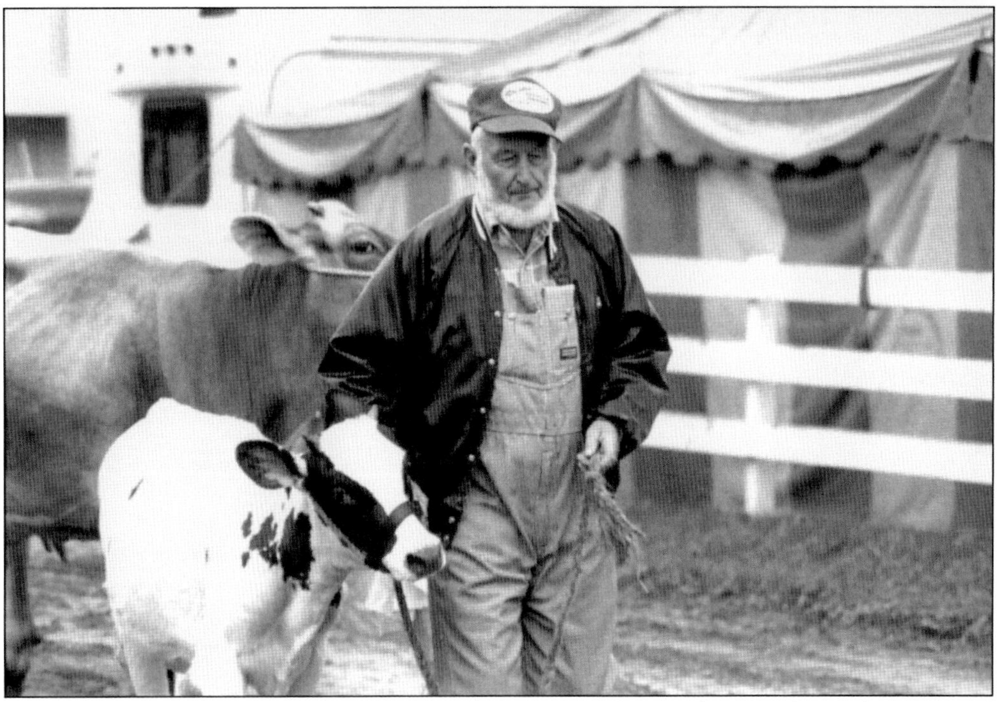

Ed Conant, a Randolph Center farmer, leads one of the family's fine Ayrshire calves. Several generations of the Conant family have shown their animals at the fair, having taken the top award twice.

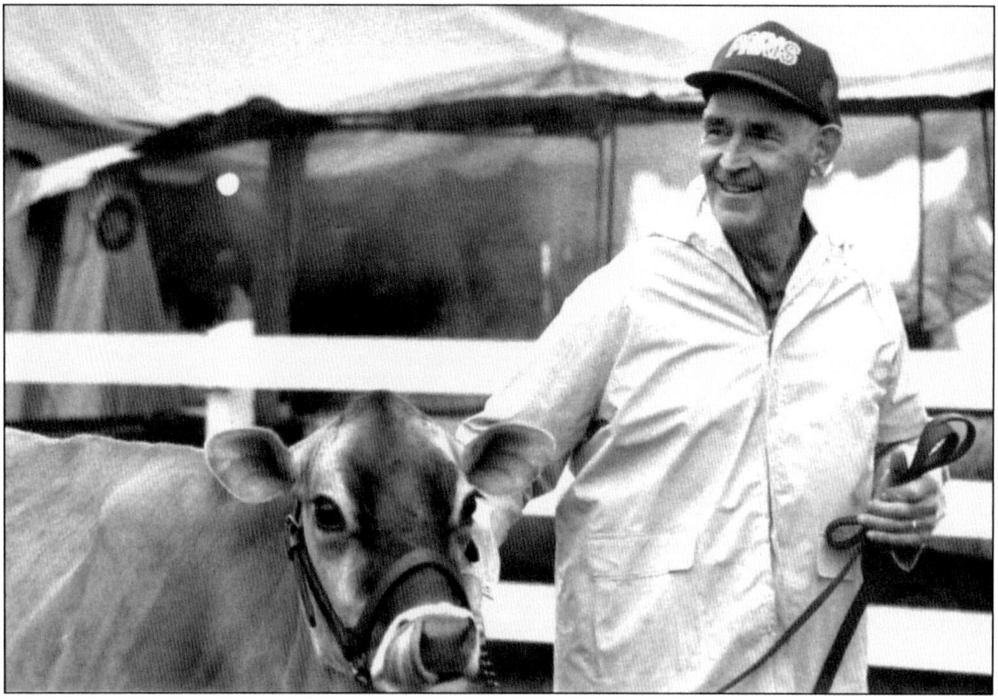

One of the best supporters of the fair has been Harold "Tuffy" Wright of Maplerow Farm in White River Junction. His Jerseys won the James K. Howe award in both 1993 and 2003. Maplerow now sponsors a trophy for the grand champion Jersey cow. The first winner was Joe Angell in 2005.

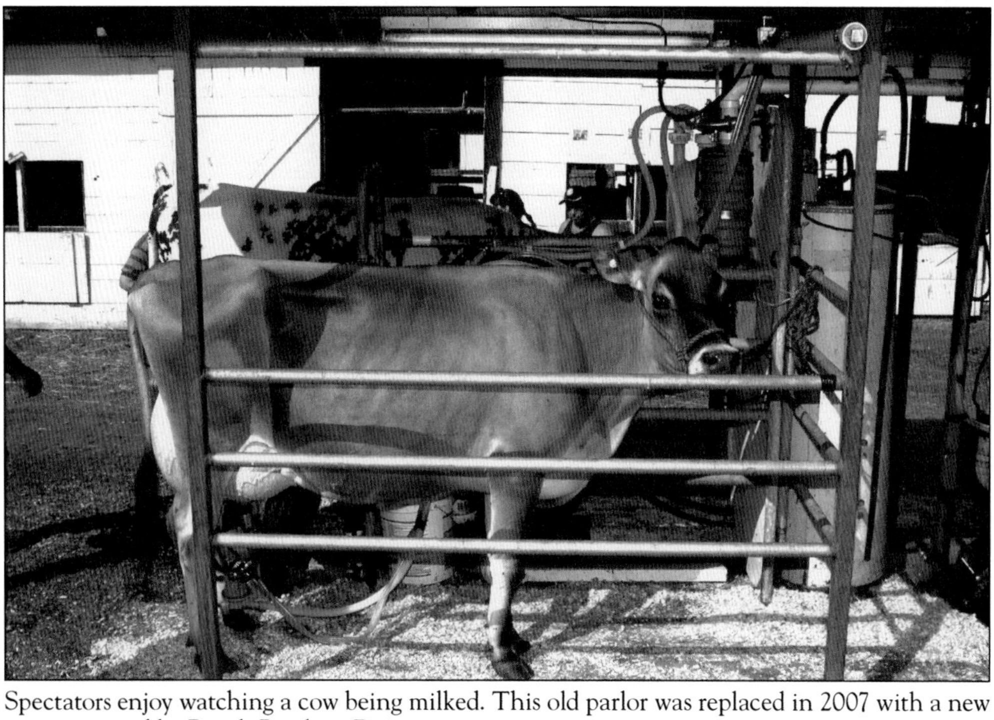

Spectators enjoy watching a cow being milked. This old parlor was replaced in 2007 with a new one sponsored by Booth Brothers Dairy.

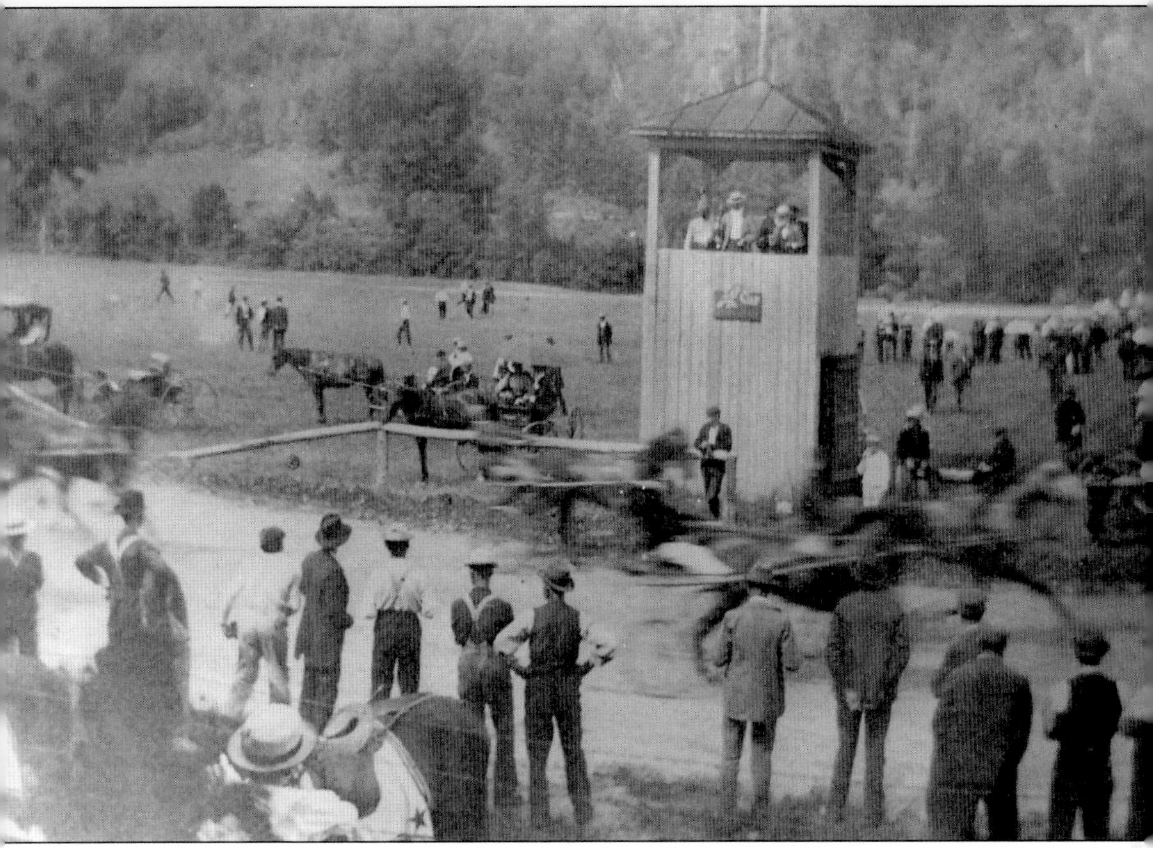

The 1898 photographer was not able to focus properly on the horses and sulkies as they streaked by him (across from the judges' stand). Liability concerns were not an issue in those days, illustrated by the spectators lining the racetrack.

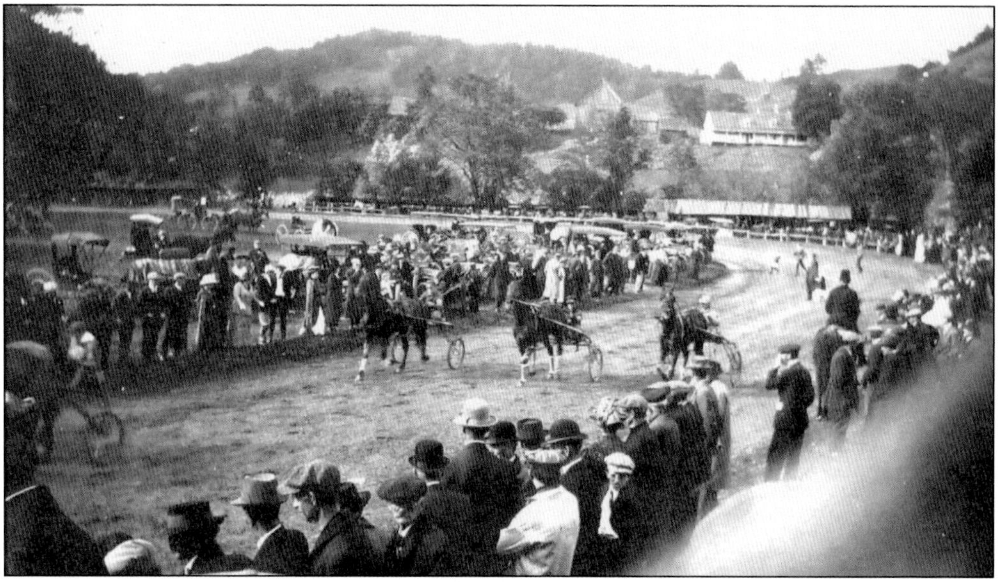

Harness racing was extremely popular in the early days of the fair, as can be seen in this 1912 photograph. Notice the crowd lining the track as the drivers are preparing to begin. On the right the mounted gentleman wearing a top hat is the race marshal.

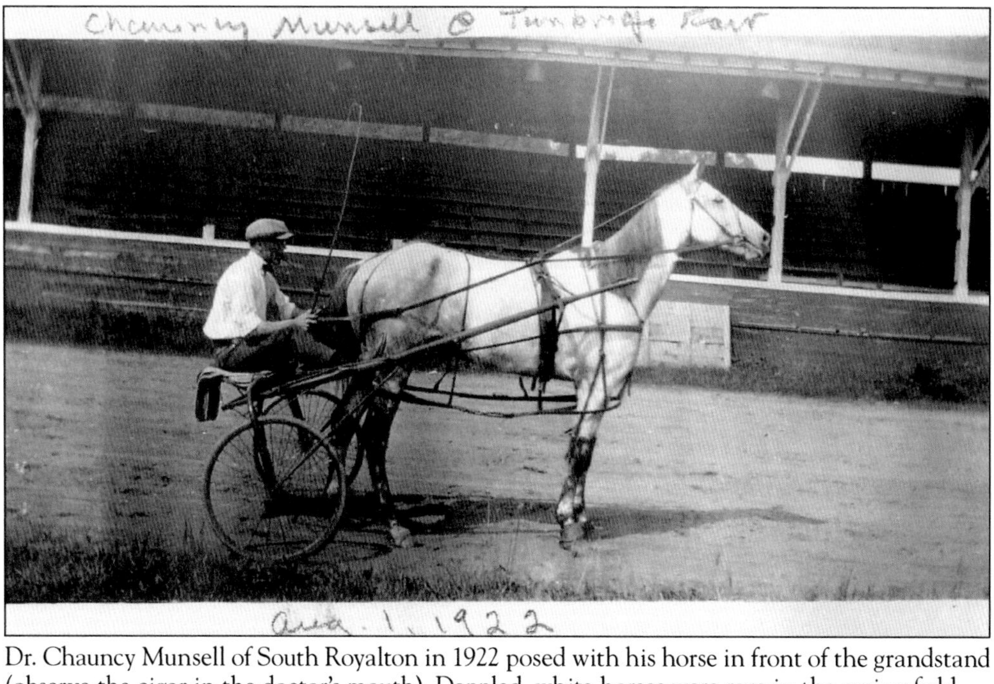

Dr. Chauncy Munsell of South Royalton in 1922 posed with his horse in front of the grandstand (observe the cigar in the doctor's mouth). Dappled, white horses were rare in the racing field.

Perhaps the Phipard family attended the 1922 fair to watch Dr. Munsell, their uncle, in a sulky race. The young boy in the touring car is Charles Phipard and the girl with the large bow in her hair is Helen (Phipard) Dumville.

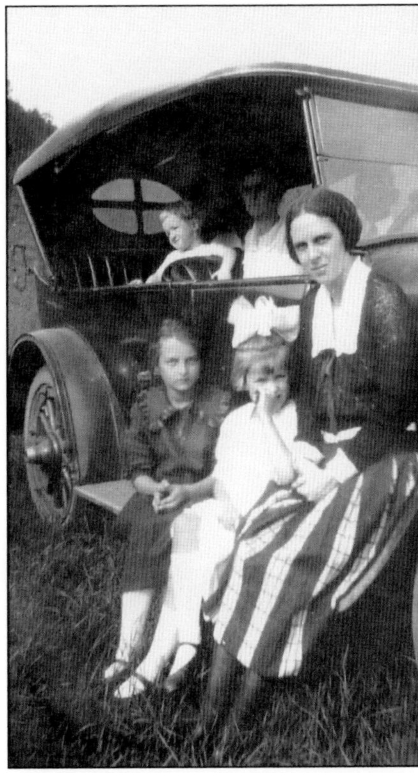

ENTRY BLANK

....1931

C. K. SMITH,
Supt. of Races,
East Bethel, Vt.

Please enter in Class ....

Name of Horse ....

Color and Sex ....

Sire ....

Sire of Dam ....

Driver's Colors ....

Owner of Horse ....

Entered by ....

Amount enclosed $ ....

Signature ....

P. O. Address ....

Fill out this blank with ink.

RACE PROGRAM
and
ENTRY BLANK

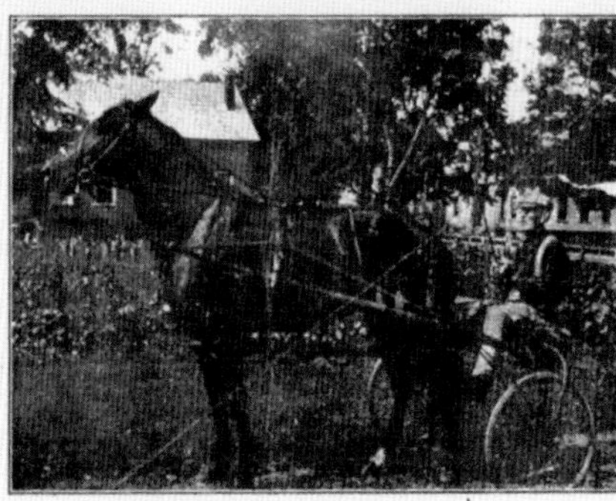

BERTONE - D. M. ROBERTS "UP"

The World's Fair
Tunbridge, Vermont
SEPTEMBER 22, 23 and 24, 1931

The 1931 racing program featured Tunbridge resident Dan M. Roberts and his horse Bertone. Sulky racing was a big event at the fair in those days. The prize money for each race was $150, a lot of money during the Depression. (Courtesy of Marie Danforth.)

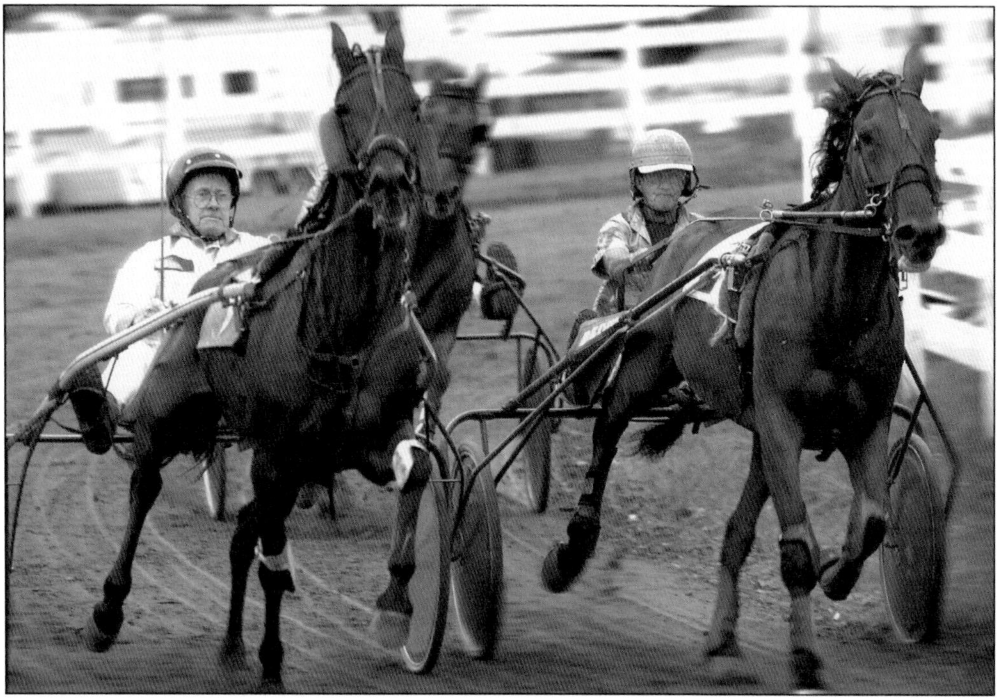

An exciting moment frozen-in-time pictures two veteran drivers rounding the north curve at the Tunbridge trotting park. Warren "Whitey" White, left, simultaneously held the records for both the trot and the pace at the fair. Ken Martin, right, in 1953 began a racing career that lasted over 50 years. He was the leading driver at Tunbridge seven times. (Courtesy of Dave White.)

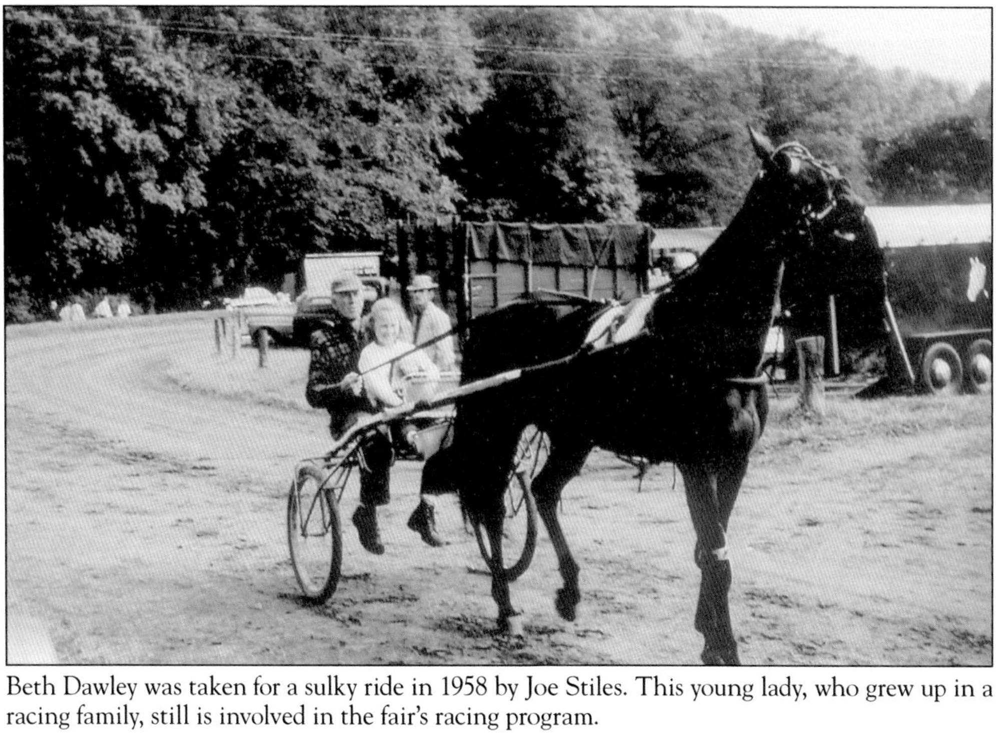

Beth Dawley was taken for a sulky ride in 1958 by Joe Stiles. This young lady, who grew up in a racing family, still is involved in the fair's racing program.

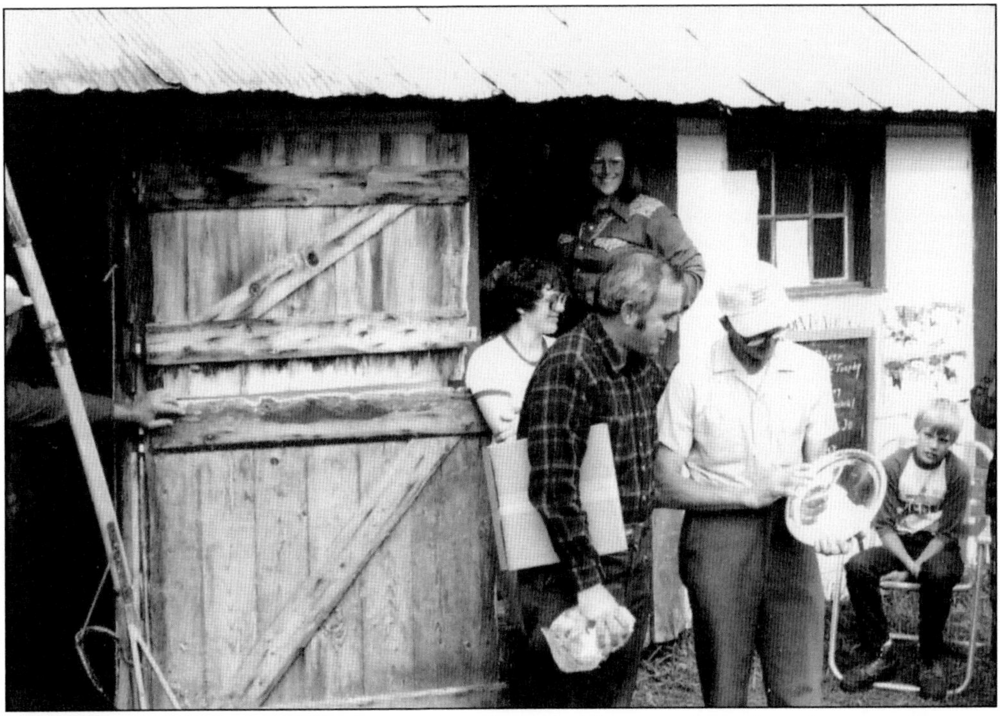

Henry Dodge, a former race superintendent for over 50 years, showed a silver award tray to Nancy Bascom (Howe), on the left, Beth Dawley, and Peter Langrock in front of the race superintendent's office.

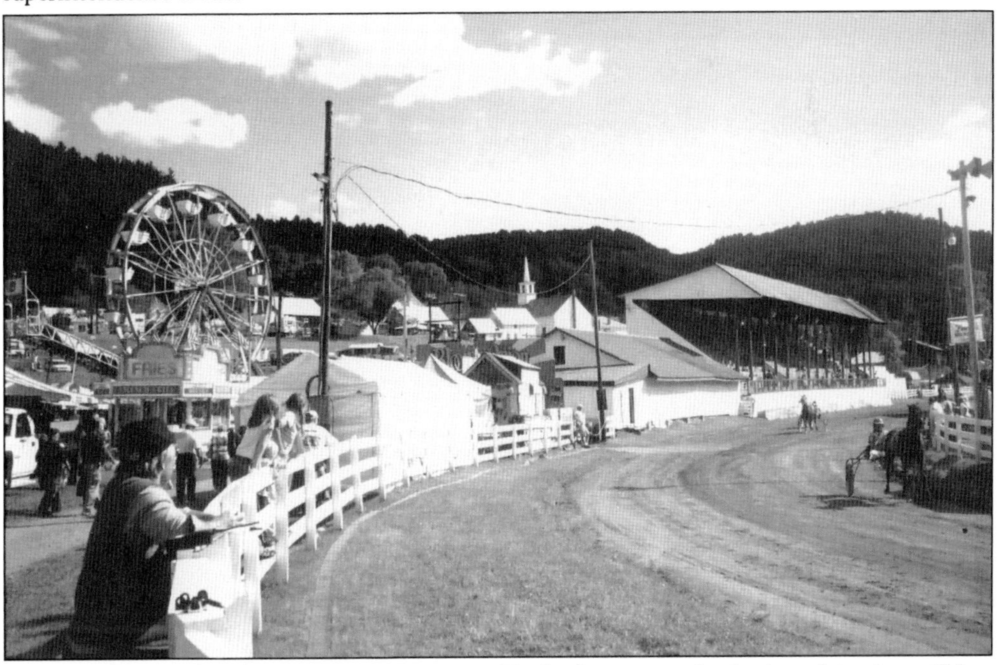

Nancy Howe stands at her post as a race passes. Tunbridge is one of only three fairs that still has sulky racing in Vermont. Much of the credit goes to Howe for her diligence in maintaining the program. (Courtesy of Margaret Greene.)

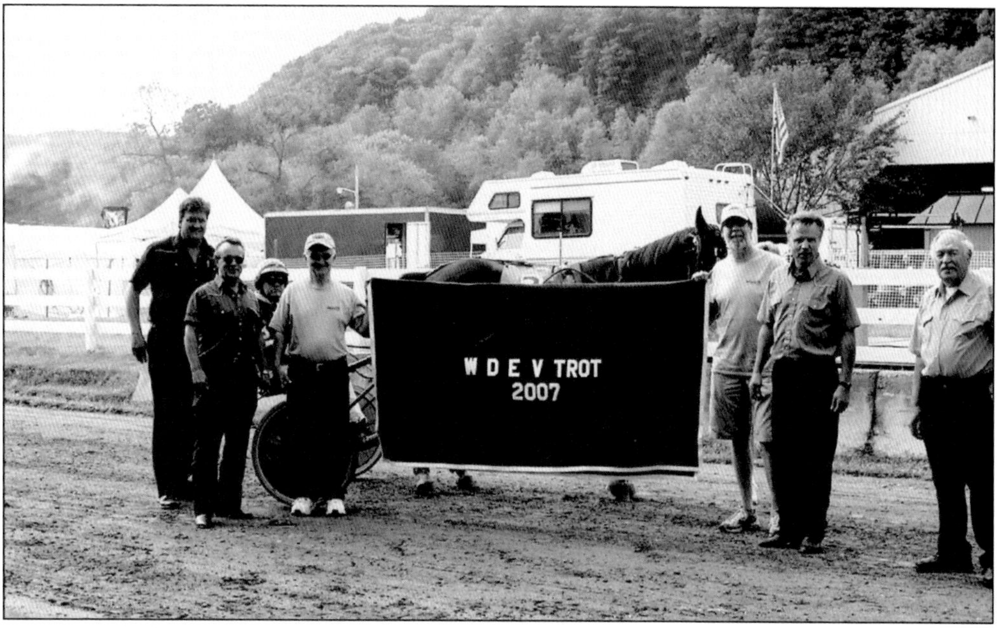

WDEV Radio has been one of the main sponsors of fair racing in recent years. Clyde Burgess Jr., 70 years old and the 2007 victor of the WDEV trot, received the winner's blanket from station personnel and the Radio Rangers. (Courtesy of Beth Dawley.)

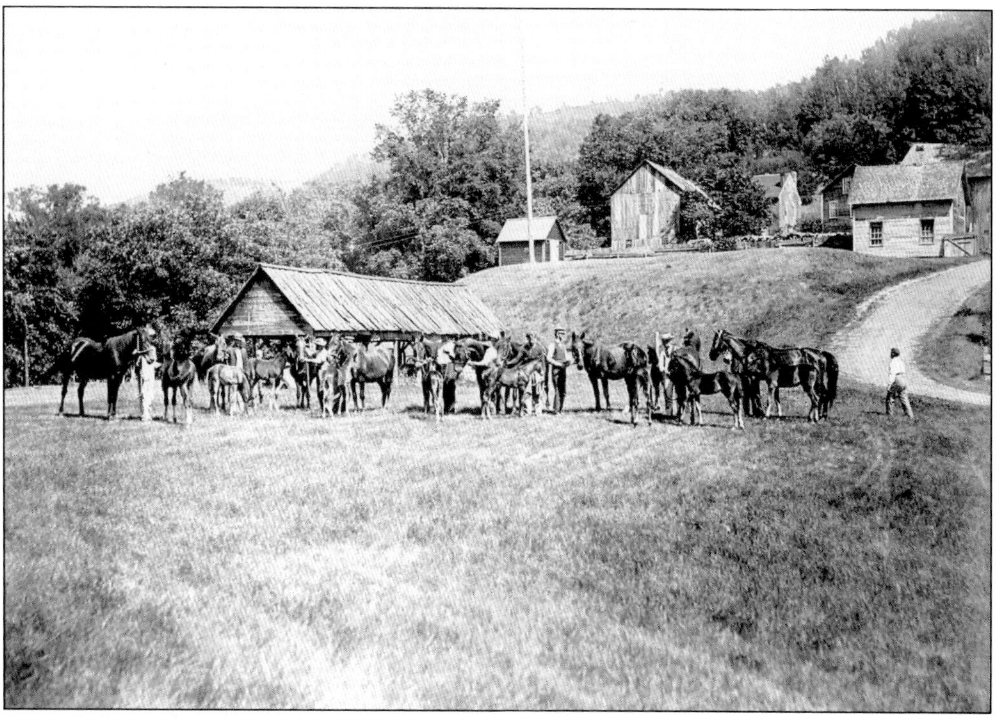

An early horse show held at the fair featured only Vermont Morgans. At the top of the road is the Main Office, which was one of the first buildings brought to the fairgrounds.

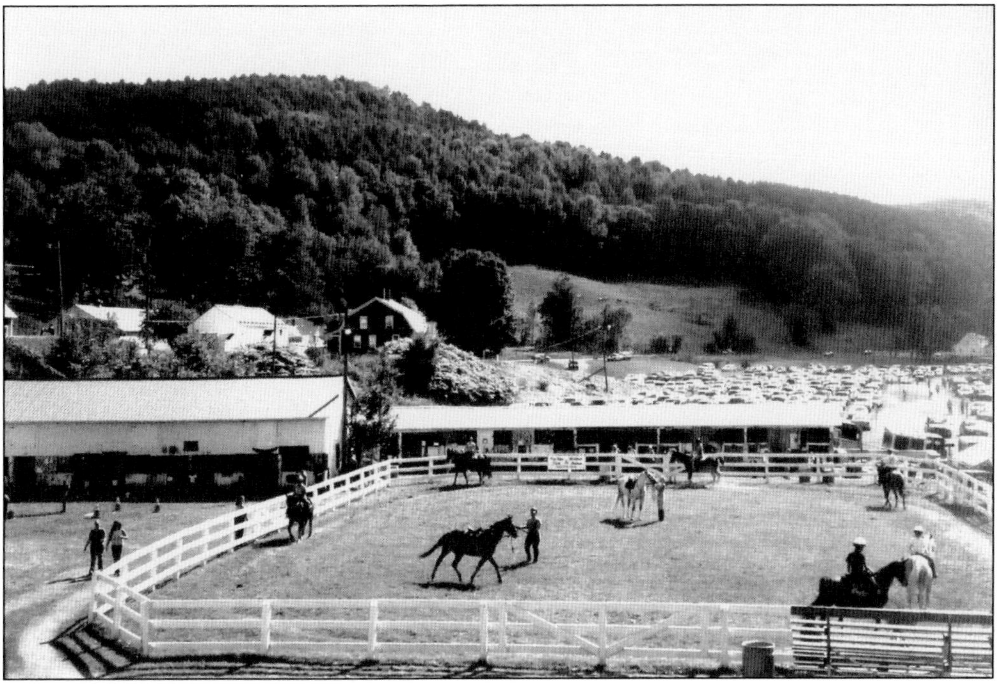

The horse show barns (horse ring in foreground), located at the south end of the grounds, are full each year with waiting lists. (Courtesy of Margaret Greene.)

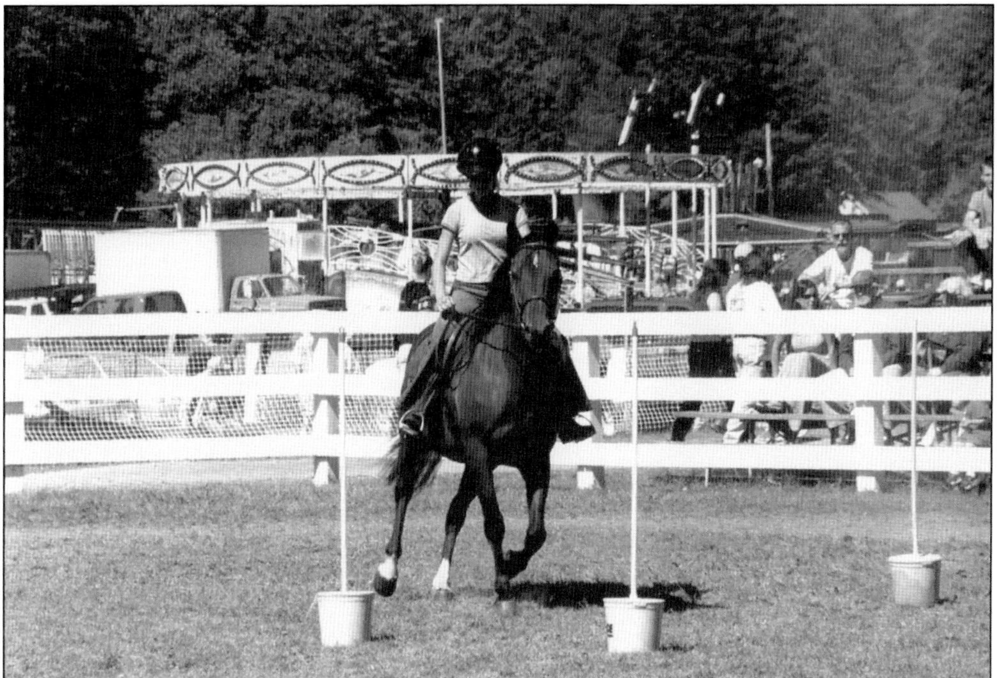

At the show ring on Sunday of the fair week a gymkhana is held as a fitting end to a long week of horse showing for the participants.

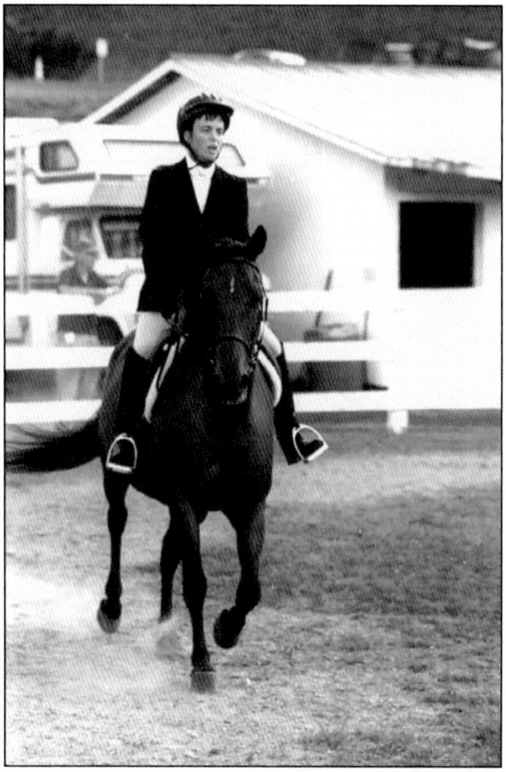

Nancy Harvey rode English on her beautiful Vermont Morgan, American Maid. People of all ages take their riding seriously. (Courtesy of Nancy Cassidy.)

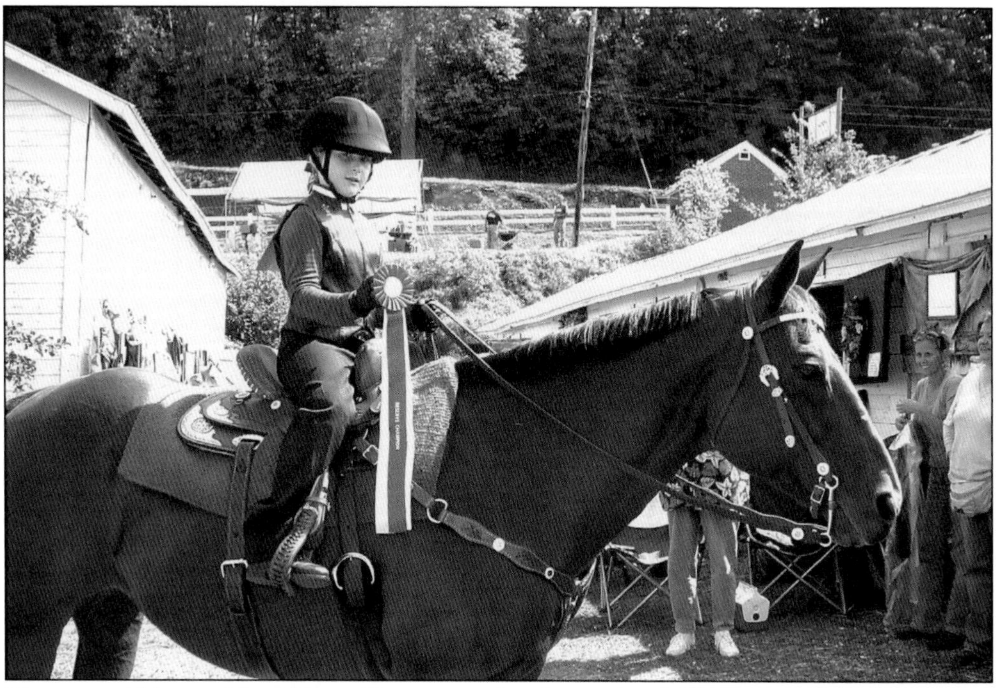

Nine-year-old Logan Metcalf proudly poses on her horse, Jewel. She was reserve champion in Western equitation in 2007. Metcalf has competed at the fair since she was five. (Courtesy of the Metcalf family.)

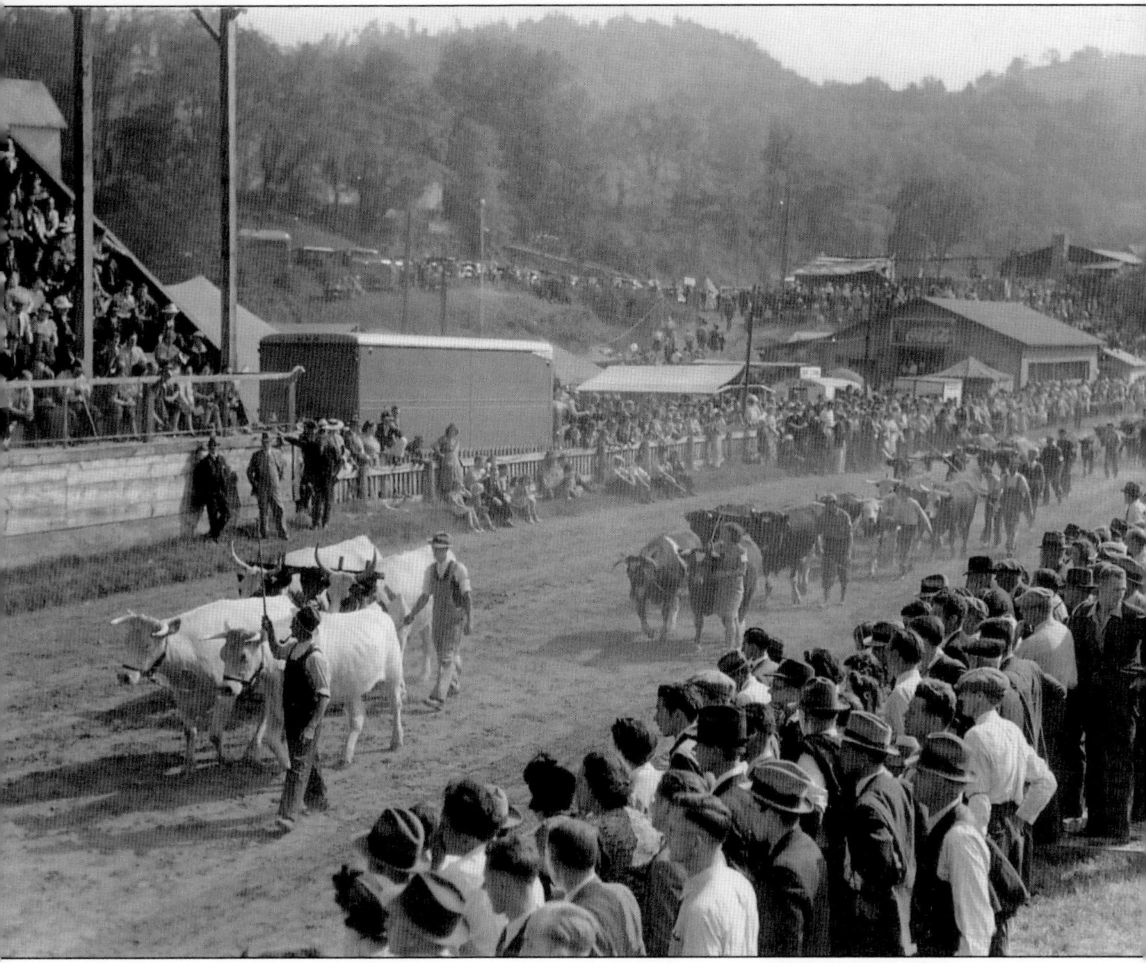

The cavalcade of animals has been one of the most popular events held on Saturday each year (weather permitting). Clifford Scoffield extensively photographed the Tunbridge Fair in 1941. He nicely illustrated a jammed grandstand, spectators crowded in front and behind the fence and observers lined up on the brow of Antique Hill. The building with the "Coca Cola" sign on the front was the popular Dreamland Dance Hall.

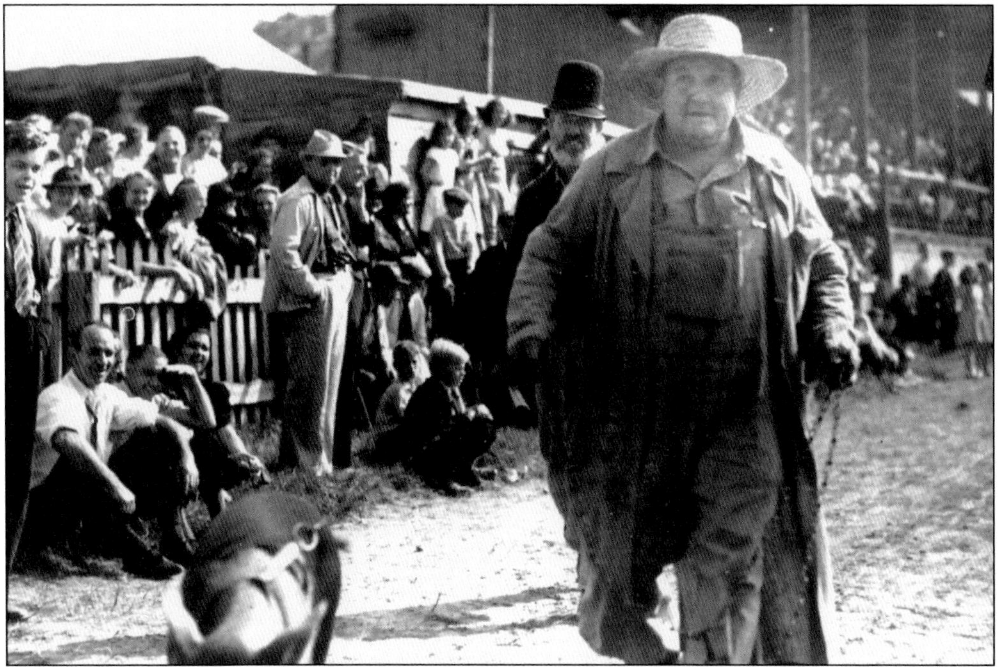

Edward Flint, the fair's secretary 1920–1951, was an indomitable showman who loved to lead his pet pig in the cavalcade in front of a packed grandstand. His uniform during the fair was always overalls, a long duster, and a wide-brimmed straw hat. Notice the "Keystone Cop" right behind him.

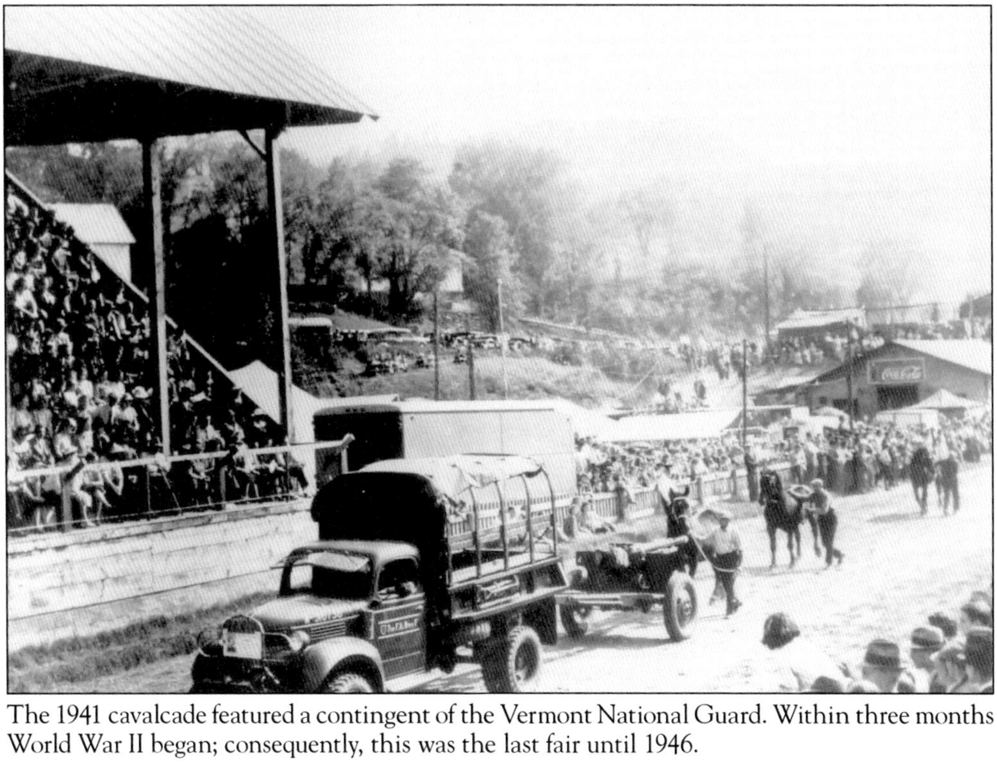

The 1941 cavalcade featured a contingent of the Vermont National Guard. Within three months World War II began; consequently, this was the last fair until 1946.

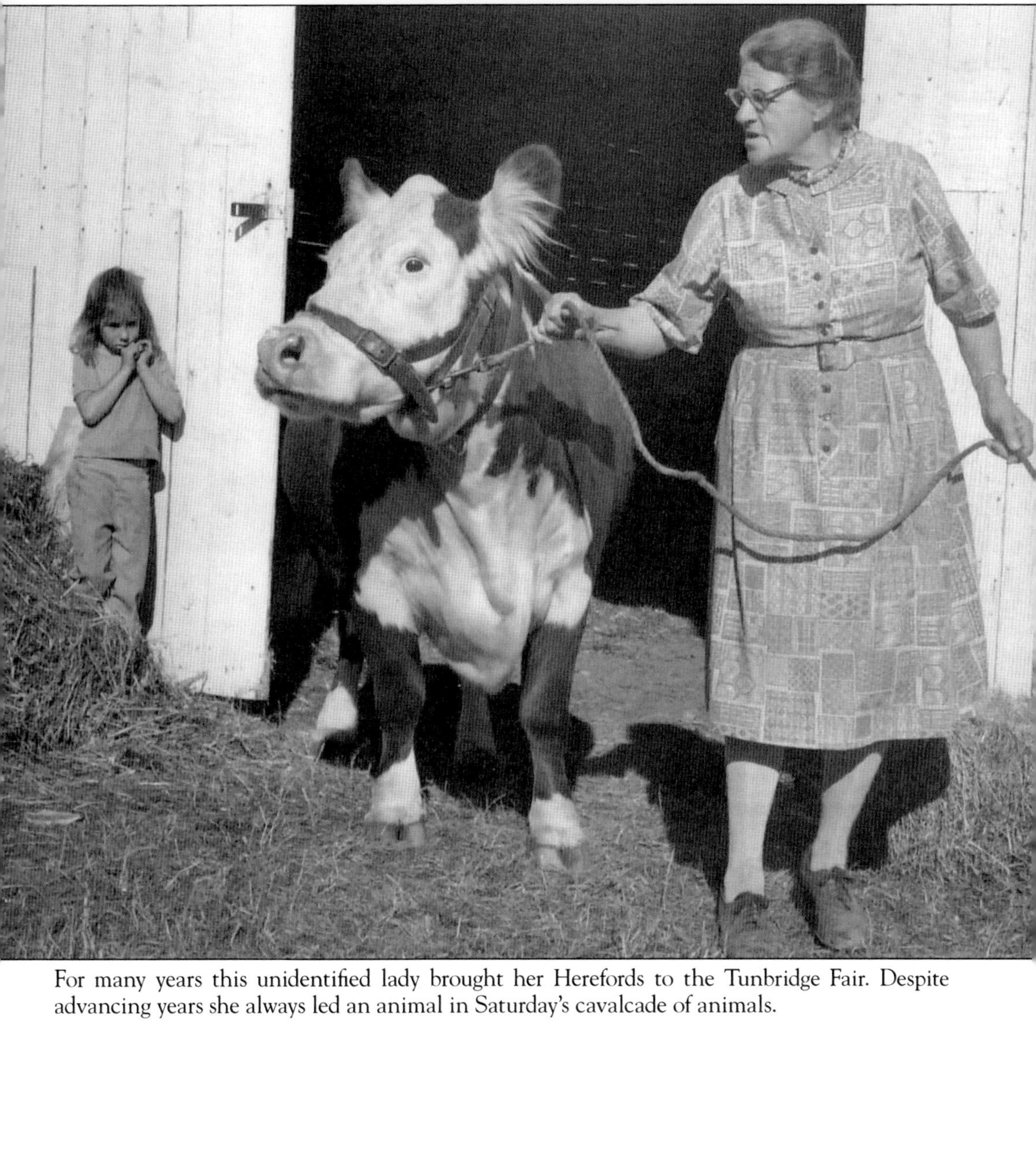

For many years this unidentified lady brought her Herefords to the Tunbridge Fair. Despite advancing years she always led an animal in Saturday's cavalcade of animals.

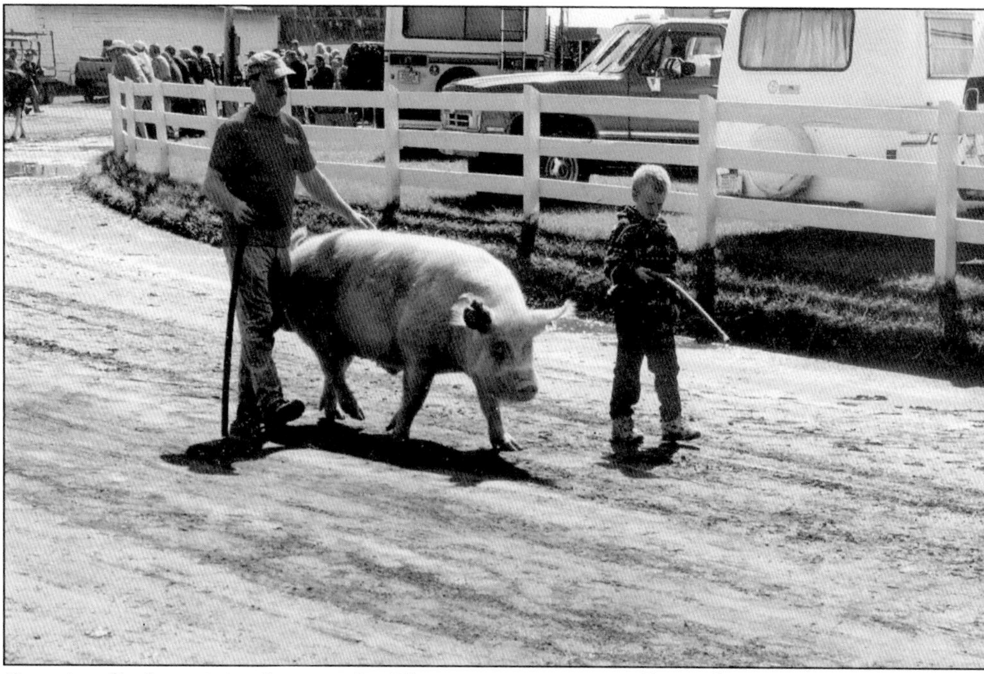

Occasionally hogs join the parade. This participant is strolling along without the need of a tether.

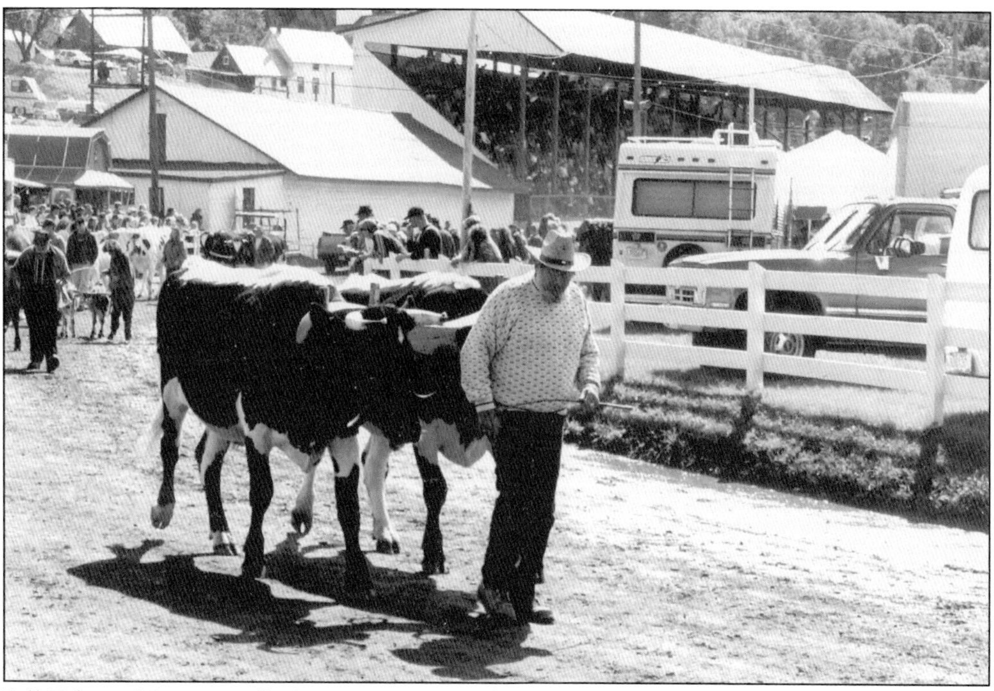

Bill Tabor of Cuttingsville, Vermont, is one of the fair's most vocal boosters and a fixture at Tunbridge's ox show. Tabor's handsome yoke of oxen is exceptionally well-trained.

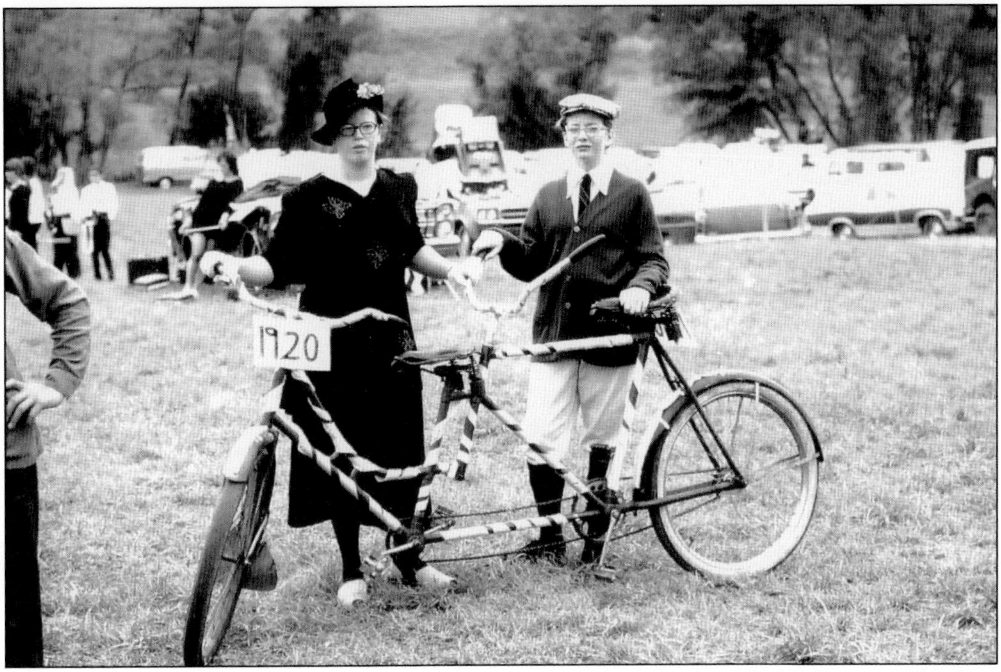

The Union Agricultural Society celebrated its centennial Tunbridge World's Fair in 1971. One of the added attractions was a special parade led by Dean C. Davis, the governor of Vermont. Susan Cilley (Barnaby) and Patricia Cilley (Ladd) rode a bicycle built for two.

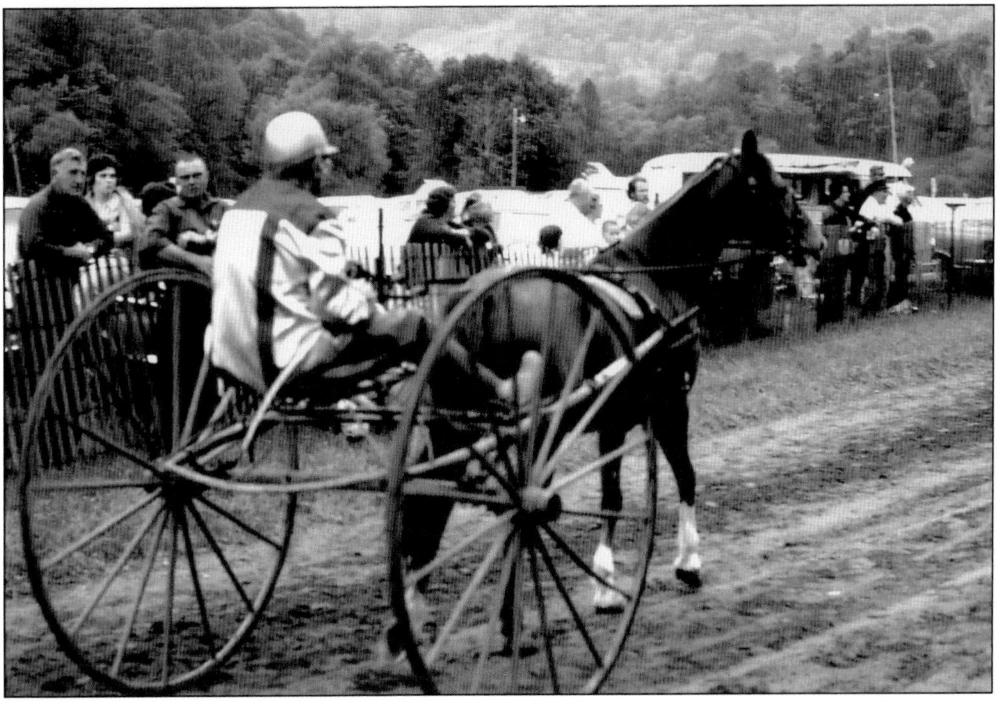

This high-wheeled sulky, which was used for racing before the 1900 fair, was stored for decades in Earl Hutchinson's barn (just south of the fairgrounds). It was borrowed for the centennial parade. The racehorse was nervous about the large wheels, but all went well.

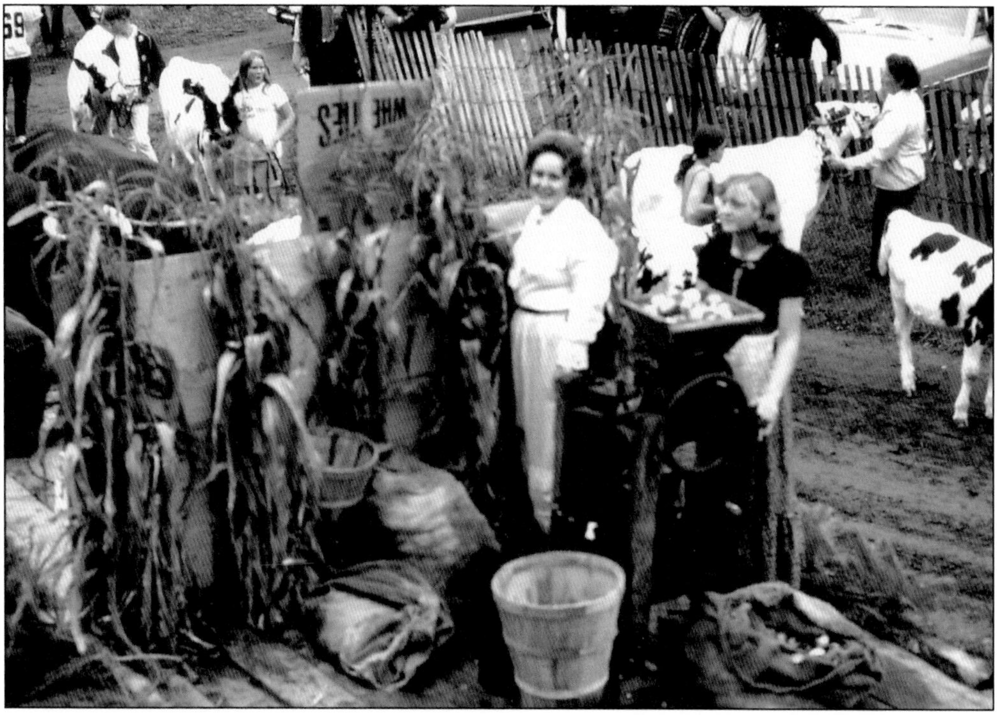

Numerous floats were entered in the huge centennial parade. Sandra (Redlon) Higgins and Sheila Metcalf rode on one that depicted a fall harvest.

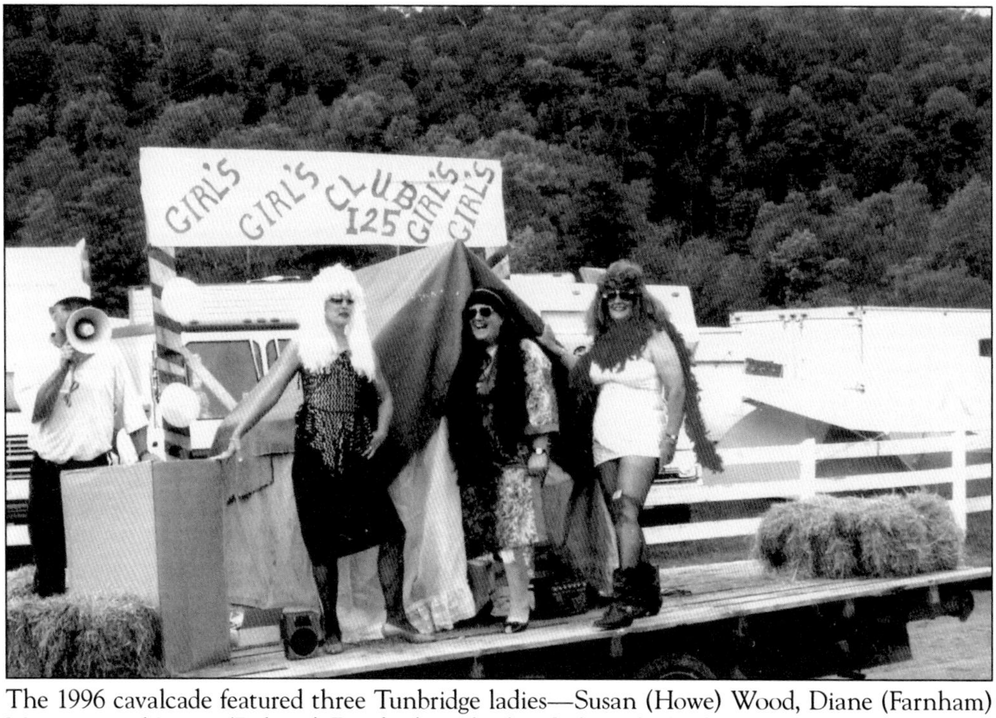

The 1996 cavalcade featured three Tunbridge ladies—Susan (Howe) Wood, Diane (Farnham) Mattoon, and Marie (Roberts) Danforth—who briefly brought back "girlie shows." This scene demonstrates why the fair banned the "girlies."

Four

Displays and Special Events

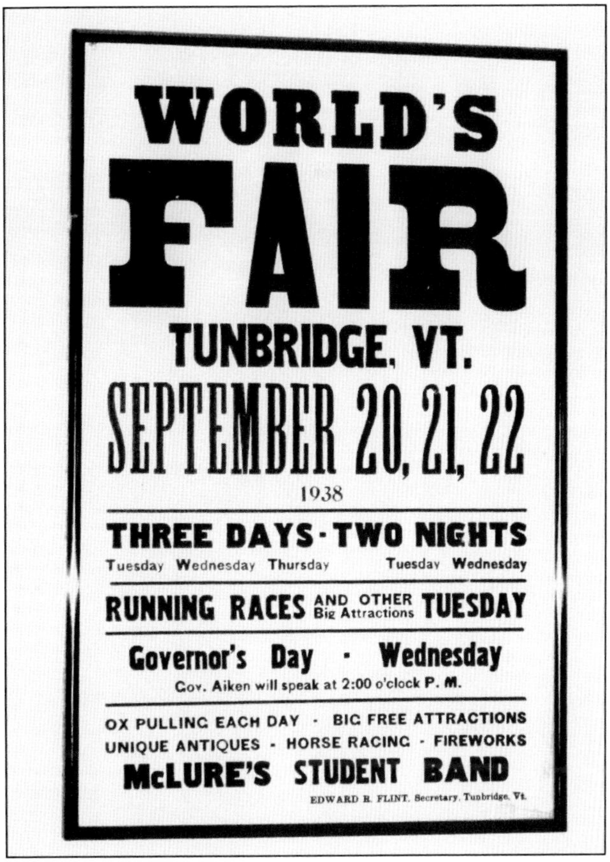

The 1938 fair poster listed the many attractions that would take place during the fair. Governor's Day was a special occasion for decades. Gov. George Aiken, one of Vermont's most beloved politicians, came to visit and speak on Wednesday.

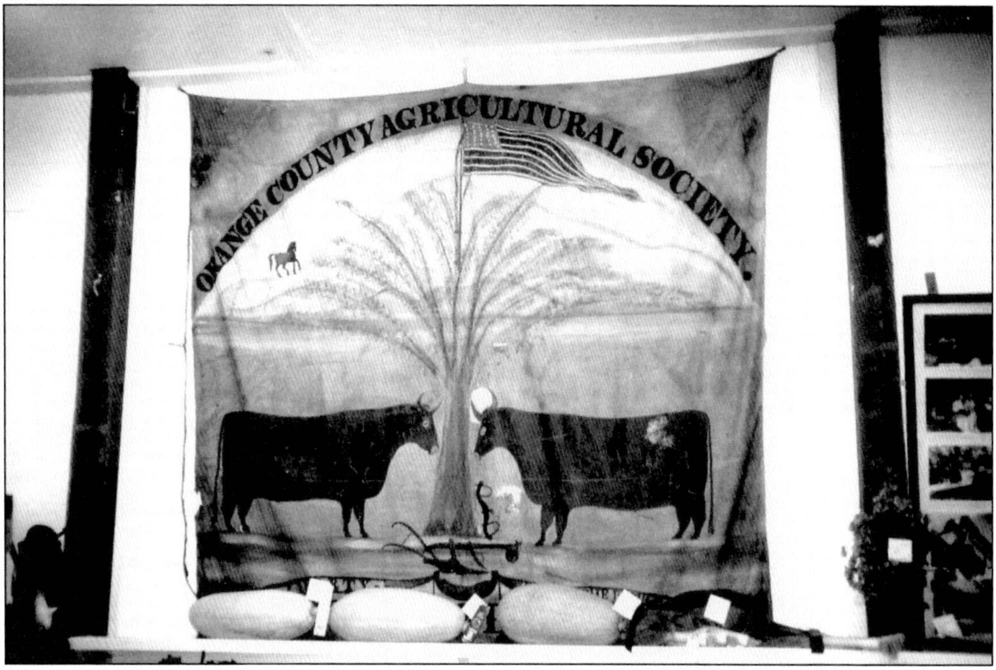

The forerunner of advertisement posters was a cloth banner. The 1858 Orange County Agricultural Society event (the predecessor of the Tunbridge Fair) banner is displayed on the wall at Dodge-Gilman Building overlooking vegetable entries. It was discovered rolled up in a dark attic, thus remarkably preserving its bright colors.

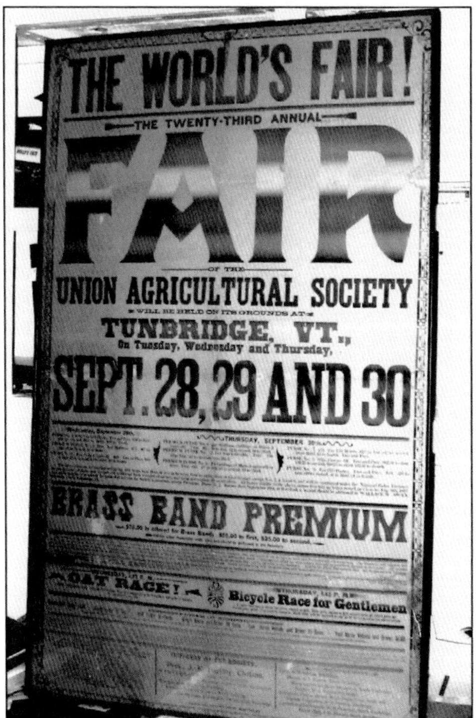

The oldest known fair poster is from 1894, the 23rd year. The featured events included a brass band competition, a bicycle race "for gentlemen," and an oat race for horses. Such a large poster would have been seen easily on the side of a barn or inside a covered bridge.

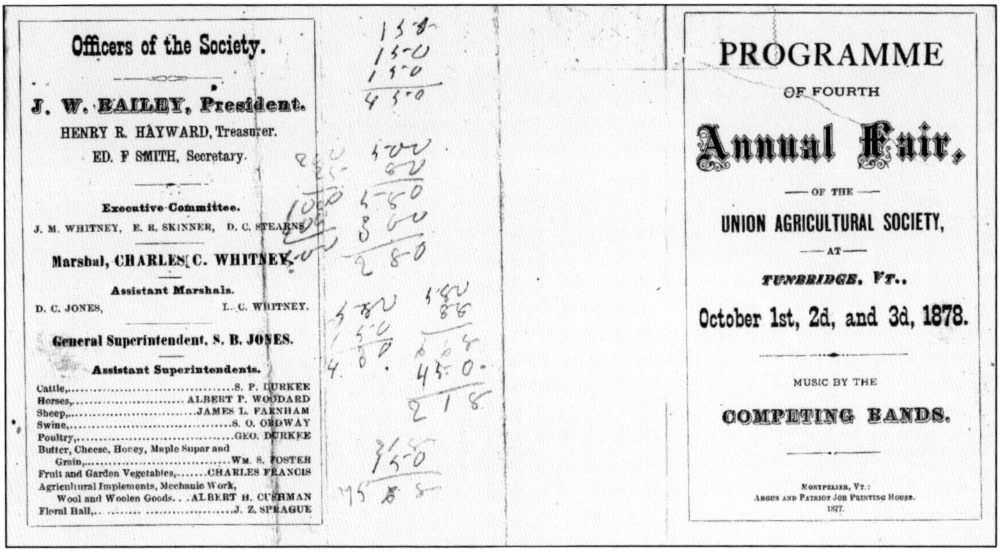
Officers of the Society.

J. W. BAILEY, President.
HENRY R. HAYWARD, Treasurer.
ED. F SMITH, Secretary.

Executive Committee.
J. M. WHITNEY, E. R. SKINNER, D. C. STEARNS

Marshal, CHARLES C. WHITNEY.

Assistant Marshals.
D. C. JONES, L. C. WHITNEY.

General Superintendent, S. B. JONES.

Assistant Superintendents.
Cattle,....S. P. DURKEE
Horses,....ALBERT P. WOODARD
Sheep,....JAMES L. FARNHAM
Swine,....S. O. ORDWAY
Poultry,....GEO. DURKEE
Butter, Cheese, Honey, Maple Sugar and Grain,....WM. S. FOSTER
Fruit and Garden Vegetables,....CHARLES FRANCIS
Agricultural Implements, Mechanic Work, Wool and Woolen Goods....ALBERT H. CUSHMAN
Floral Hall,....J. Z. SPRAGUE

PROGRAMME
OF FOURTH
Annual Fair,
OF THE
UNION AGRICULTURAL SOCIETY,
AT
TUNBRIDGE, VT.,
October 1st, 2d, and 3d, 1878.

MUSIC BY THE
COMPETING BANDS.

MONTPELIER, VT.:
ARGUS AND PATRIOT JOB PRINTING HOUSE.
1877.

The oldest known "programme" is from 1878. While racing was dominant, there were several other interesting things on the schedule. The second day's morning included a plowing match followed by an "exhibition of Matched and Gentlemen's Driving Horses on the track." A noon dinner preceded an annual address, which was delivered by Nathan Skinner of Plainfield. "Immediately after the address, the Judge on Music will review the bands, and his decision will be declared from the Judges' Stand by the Secretary. . . . At 3 o'cock [*sic*] lady Equestrianship" was awarded a $10 purse. The third day was devoted to sulky racing, noon dinner, and "Declaring Premiums from the Judges' stand, by the secretary." The purses were from $10 to $75. On a personal note, the author's great-grandfathers were assistant superintendents—James L. Farnham and George Durkee.

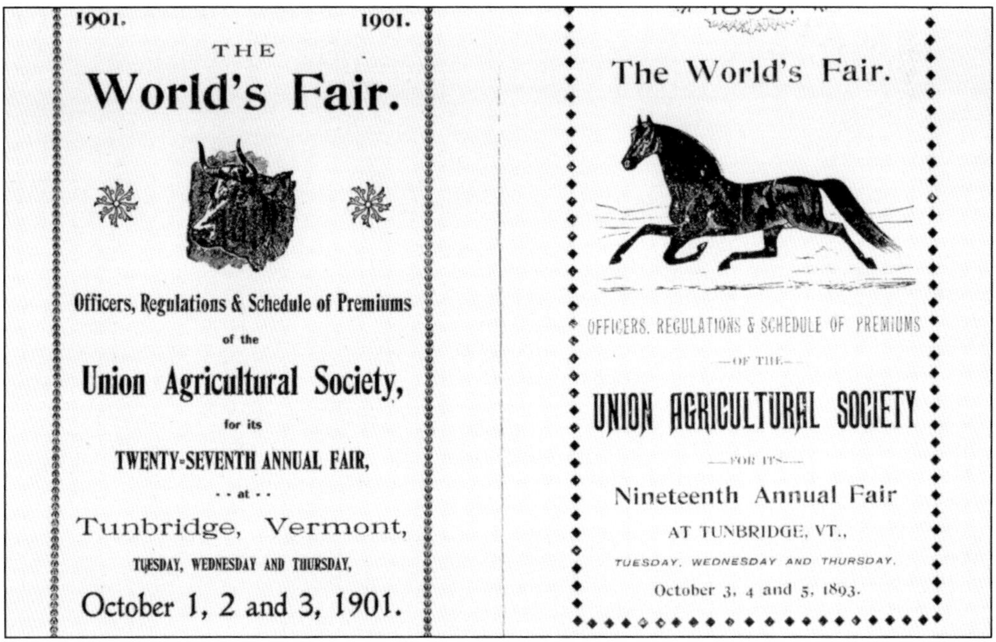

Two premium book covers from the early days of the fair depicted illustrations of the most popular events, cattle showing and sulky racing.

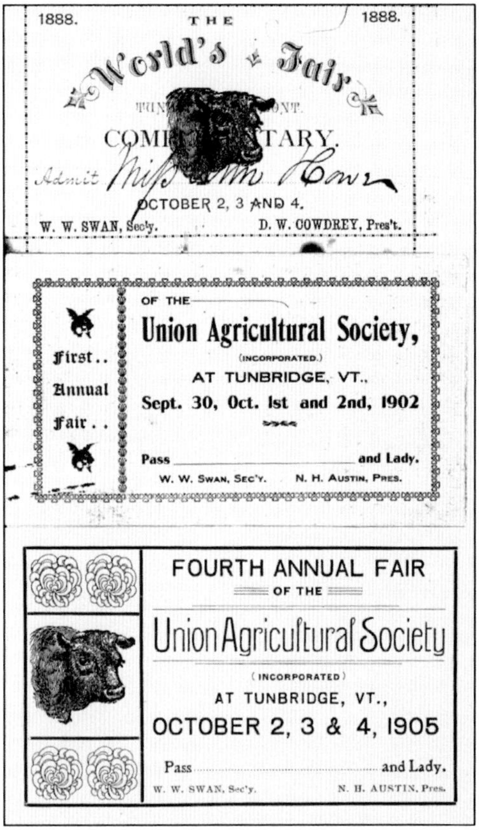

1888. THE 1888.

World's Fair

COMPLIMENTARY.

Admit

OCTOBER 2, 3 AND 4.

W. W. SWAN, Sec'y. D. W. COWDREY, Pres't.

First.. Annual Fair..

OF THE

Union Agricultural Society,

(INCORPORATED.)

AT TUNBRIDGE, VT.,

Sept. 30, Oct. 1st and 2nd, 1902

Pass ____ and Lady.

W. W. SWAN, SEC'Y. N. H. AUSTIN, PRES.

FOURTH ANNUAL FAIR

OF THE

Union Agricultural Society

(INCORPORATED)

AT TUNBRIDGE, VT.,

OCTOBER 2, 3 & 4, 1905

Pass .... and Lady.

W. W. SWAN, Sec'y. N. H. AUSTIN, Pres.

An 1888 ticket lists D. W. Cowdrey as fair president, a post he held for 12 years. He died in 1892 after which Nathaniel H. Austin took over as president (see 1902 ticket) for 15 years. Wallace Swan served as fair secretary 1888–1906.

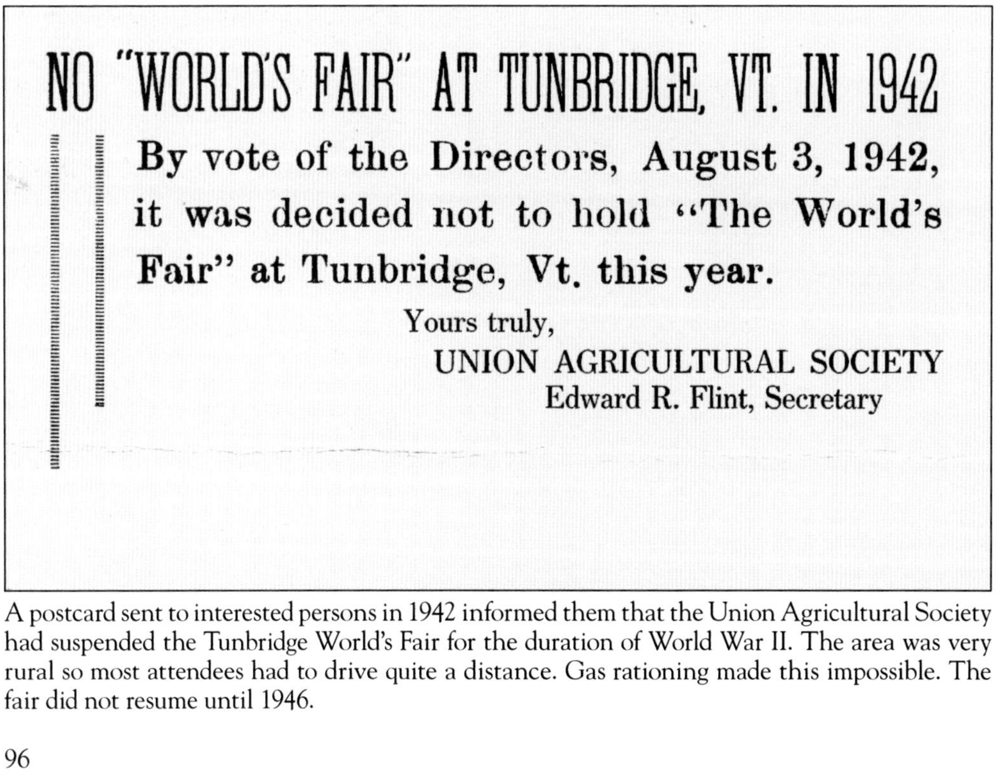

NO "WORLD'S FAIR" AT TUNBRIDGE, VT. IN 1942

By vote of the Directors, August 3, 1942, it was decided not to hold "The World's Fair" at Tunbridge, Vt. this year.

Yours truly,

UNION AGRICULTURAL SOCIETY

Edward R. Flint, Secretary

A postcard sent to interested persons in 1942 informed them that the Union Agricultural Society had suspended the Tunbridge World's Fair for the duration of World War II. The area was very rural so most attendees had to drive quite a distance. Gas rationing made this impossible. The fair did not resume until 1946.

WORLD'S FAIR

TUNBRIDGE, VT.

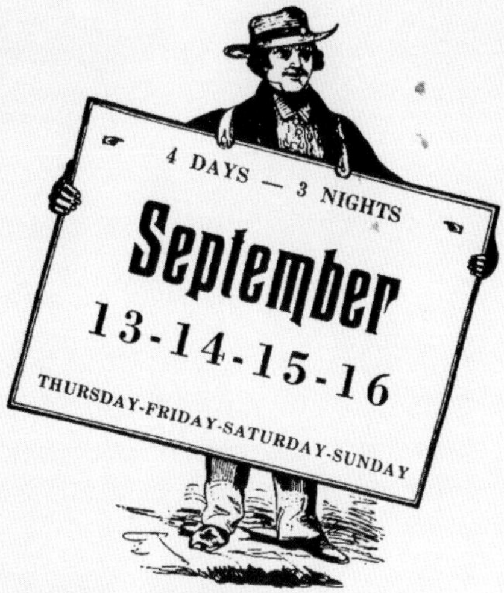

102nd ANNIVERSARY GENERAL PROGRAM

THURSDAY, SEPTEMBER 13, 1973

SPECIAL — Country Show from Nashville

9:00 a.m.	Judging begins for all exhibits
10:00 a.m.	Pony Pulling (2 classes)
1:00 p.m.	Showing of Old Type Morgan Horses
1:30 p.m.	Stage Show
2:30 p.m.	Jean Shepard & Carl Smith Show

EVENING

7:30 p.m.	Stage Show
8:00 p.m.	Jean Shepard & Carl Smith Show

FRIDAY, SEPTEMBER 14, 1973

Governor's Day

9:30 a.m.	Junior Division Livestock Show
10:00 a.m.	Horse Pulling (3 classes)
11:00 a.m.	Old Time Dancers at Log Cabin
12:30 p.m.	Horse Racing
1:00 p.m.	Stage Show begins
1:00 p.m.	Old Time Dancers at Log Cabin
1:30 p.m.	Beef Judging

EVENING

7:30 p.m.	Stage Show
8:30 p.m.	Old Time Fiddlers Contest

Antique Show — Old Time Dancers Daily
Dancing at Dreamland — Day and Evening

LISTEN TO LOUDSPEAKER FOR DAILY PROGRAMS AT CABIN.

Children's Playground, Free

Stage Show and Attractions furnished by Al Martin of Boston, Mass.

SATURDAY, SEPTEMBER 15, 1973

9:30 a.m.	Junior Division Fitting and Showing Contest
10:00 a.m.	Ox Pulling (3 classes)
11:00 a.m.	Old Time Dancers at Log Cabin
12:00 Noon	Band Concert on Stage
12:30 p.m.	Horse Racing
1:00 p.m.	Stage Show begins
1:00 p.m.	Old Time Dancers at Log Cabin
3:30 p.m.	Parade of Livestock and Machinery

EVENING

7:30 p.m.	Band Concert and Drill by McLure's Student Band
8:00 p.m.	Stage Show

SUNDAY, SEPTEMBER 16, 1973

10:00 a.m.	Fair "Festival of Worship" Service at Congregational Church (come as you are)
11:00 a.m.	Exhibits Open
12:30 p.m.	Old Time Dancers at Log Cabin
2:00 p.m.	Stage Show
3:00 p.m.	Auto Thrill Show

Admissions

Single Ticket Thursday	$1.45 + .05 Tax =	$1.50
Single Ticket Friday	1.45 + .05 Tax =	1.50
Single Ticket Saturday	1.94 + .06 Tax =	2.00
Single Ticket Sunday	1.45 + .05 Tax =	1.50
Single Seasons Ticket Adult	4.85 + .15 Tax =	5.00
Children under 12 (Free Thursday)	.48 + .02 Tax =	.50
Automobile		1.00
Motorcycle		.50
Grandstand Thursday & Friday	.97 + .03 Tax =	1.00
Grandstand Saturday & Sunday	1.45 + .05 Tax =	1.50
Reserved Grandstand	1.94 + .06 Tax =	2.00
Grandstand Evening	.97 + .03 Tax =	1.00
Seasons Ticket Camper or Camper Trailer		10.00

Autos will be allowed on the grounds but not on the track. Campers parked in special area. No tents allowed in parking area.

UNION AGRICULTURAL SOCIETY • TUNBRIDGE, VT.

Nashville country shows, a fiddlers' contest, vaudeville acts, and an automobile thrill show headlined the 1973 program. Ticket prices had increased to $1.50 and $2, and children were charged 50¢. A grandstand ticket could be purchased for $1.50.

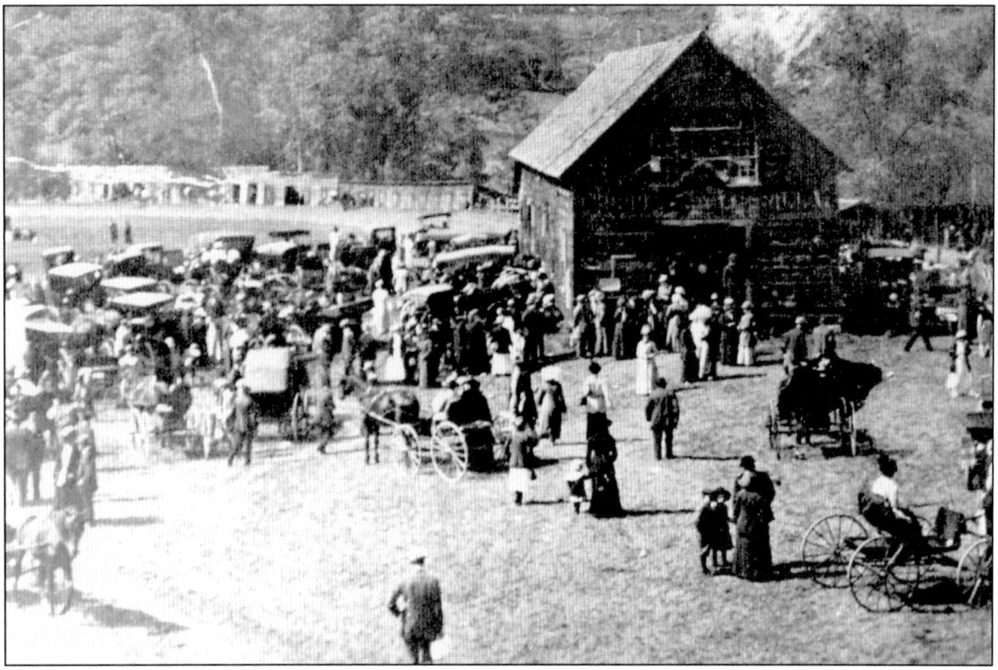

"Hood's Sarsaparilla" was painted in large letters across the front of Floral Hall, as seen faintly on a very early photograph. Harvey Hood, whose family was from Chelsea, may have contributed toward costs of Floral Hall. Exhibits include sewn items, knitting, crocheting, embroidery, hooking, canned goods, baked foods, photography, art, crafts, houseplants, and flowers. Exhibitor entries are at an all-time high.

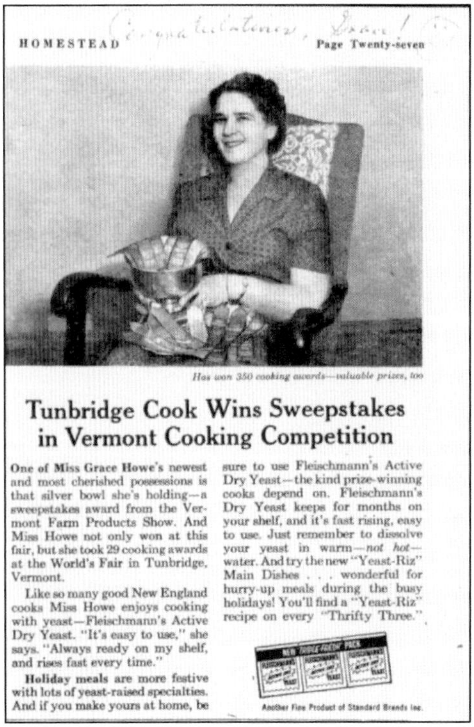

HOMESTEAD Page Twenty-seven

Has won 350 cooking awards—valuable prizes, too

Tunbridge Cook Wins Sweepstakes in Vermont Cooking Competition

One of Miss Grace Howe's newest and most cherished possessions is that silver bowl she's holding—a sweepstakes award from the Vermont Farm Products Show. And Miss Howe not only won at this fair, but she took 29 cooking awards at the World's Fair in Tunbridge, Vermont.

Like so many good New England cooks Miss Howe enjoys cooking with yeast—Fleischmann's Active Dry Yeast. "It's easy to use," she says. "Always ready on my shelf, and rises fast every time."

Holiday meals are more festive with lots of yeast-raised specialties. And if you make yours at home, be sure to use Fleischmann's Active Dry Yeast—the kind prize-winning cooks depend on. Fleischmann's Dry Yeast keeps for months on your shelf, and it's fast rising, easy to use. Just remember to dissolve your yeast in warm—*not hot*—water. And try the new "Yeast-Riz" Main Dishes . . . wonderful for hurry-up meals during the busy holidays! You'll find a "Yeast-Riz" recipe on every "Thrifty Three."

Another Fine Product of Standard Brands Inc.

Grace Howe lived her 91 years in Tunbridge, many of them entering her food products in Floral Hall. In 1957, she won an astounding 29 ribbons. The silver bowl was a sweepstakes award received at the Vermont Farm Show.

Marie Danforth, who grew up with the fair as an important part of her life, over the years has been involved in many fair activities: cleaning, assisting in the antique museum, and being an important part of Floral Hall. In addition she has earned many blue ribbons with her baking (especially maple) at the Tunbridge Fair and the Vermont Farm Show. (Courtesy of Nancy Cassidy.)

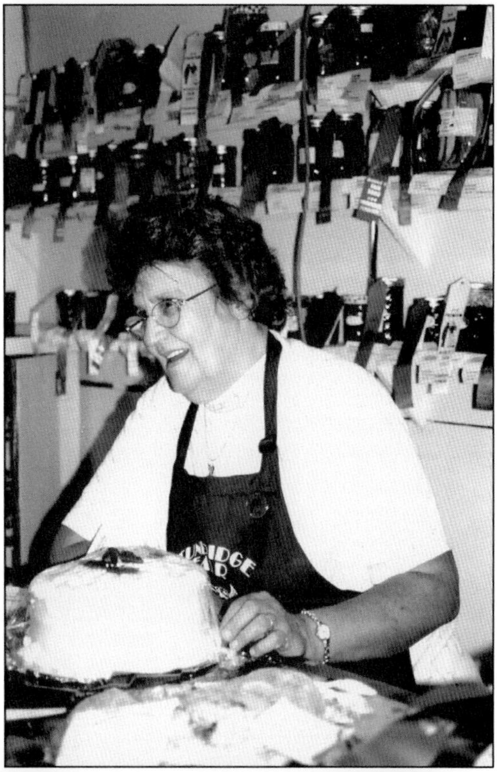

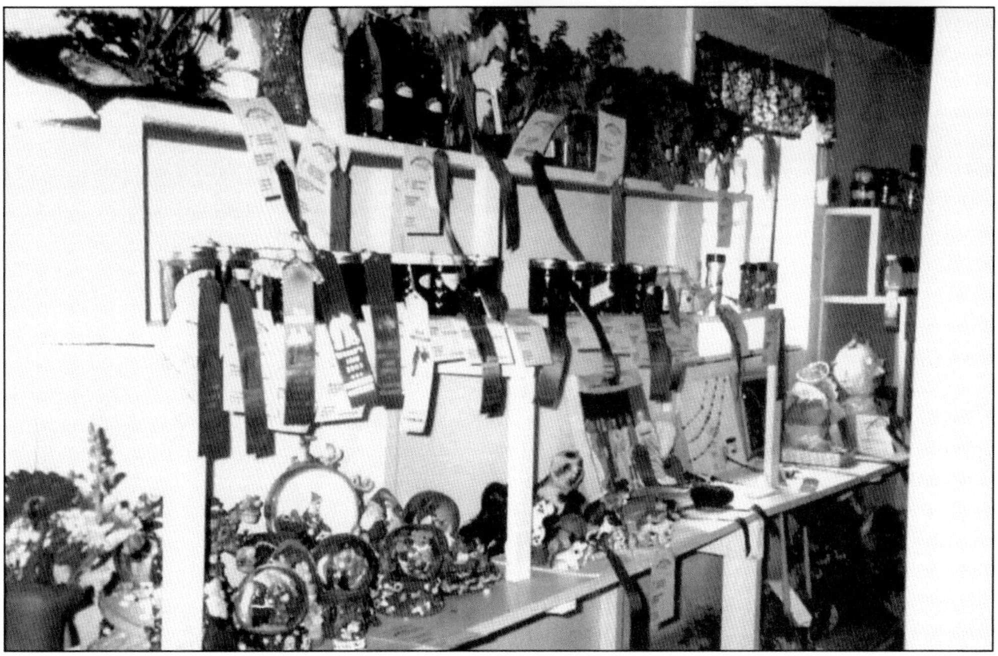

The junior division is "bursting at the seams" with all kinds of projects from young people. So many children proudly bring their items yearly to the hall that space has become a critical problem. Tunbridge's terminology—youth division, rather than 4-H club—enables all young people to enter the categories. (Courtesy of Nancy Cassidy.)

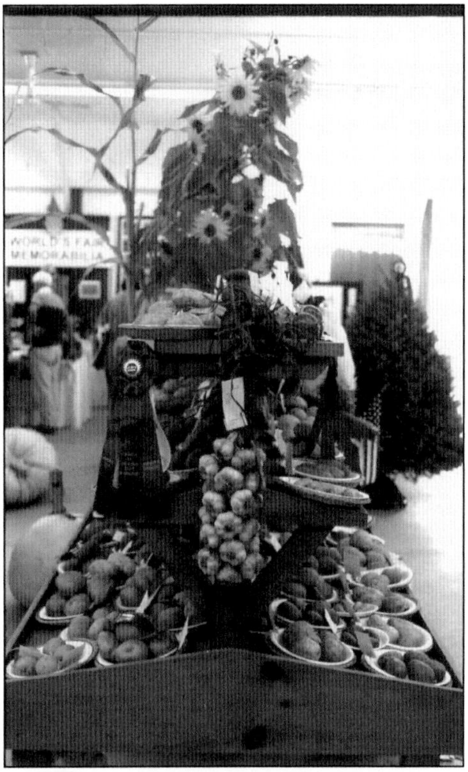

In the early years, hops were an important crop judged at the fair; however, it has been a long time since any hops were displayed. Dodge-Gilman houses entries of fruits, vegetables, hay, grains, the tallest sunflower, and large pumpkins. (Courtesy of Nicholas Wolff.)

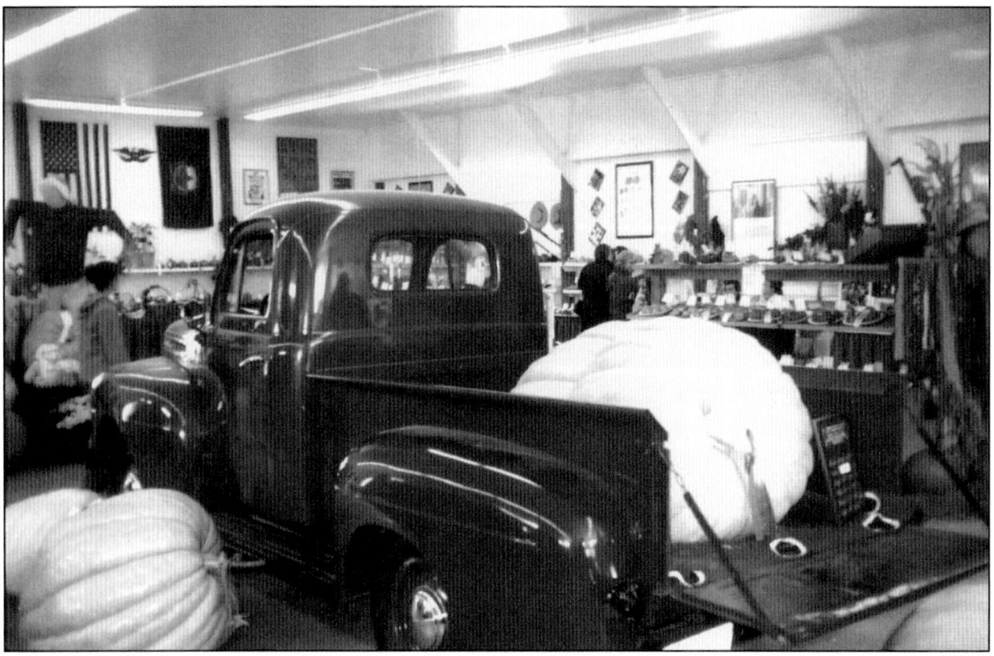

Everyone wants to see the big pumpkins in Dodge-Gilman. Each year the superintendent tries to vary the display. A few men and boys yearly vie for the award in one of the most competitive events. The 2007 biggest pumpkin, grown by Dr. Michael Curtis of South Strafford, Vermont, weighed in at 718.5 pounds.

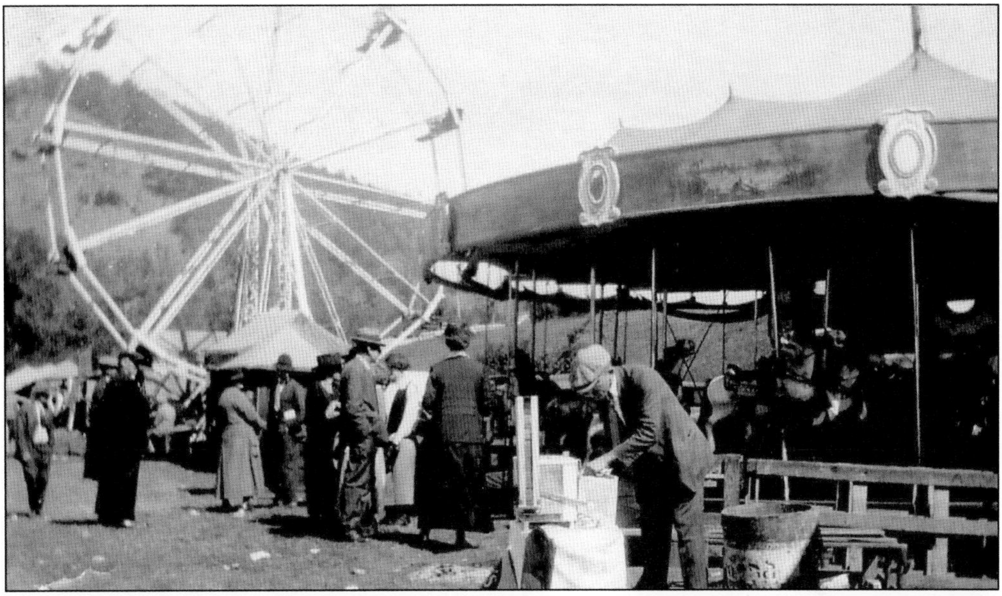

It is uncertain when the first carnival arrived at the fair; however, there is no question that the merry-go-round was the only ride for several years. Later it was joined by a Ferris wheel, as seen in a view of the midway in the early 1930s.

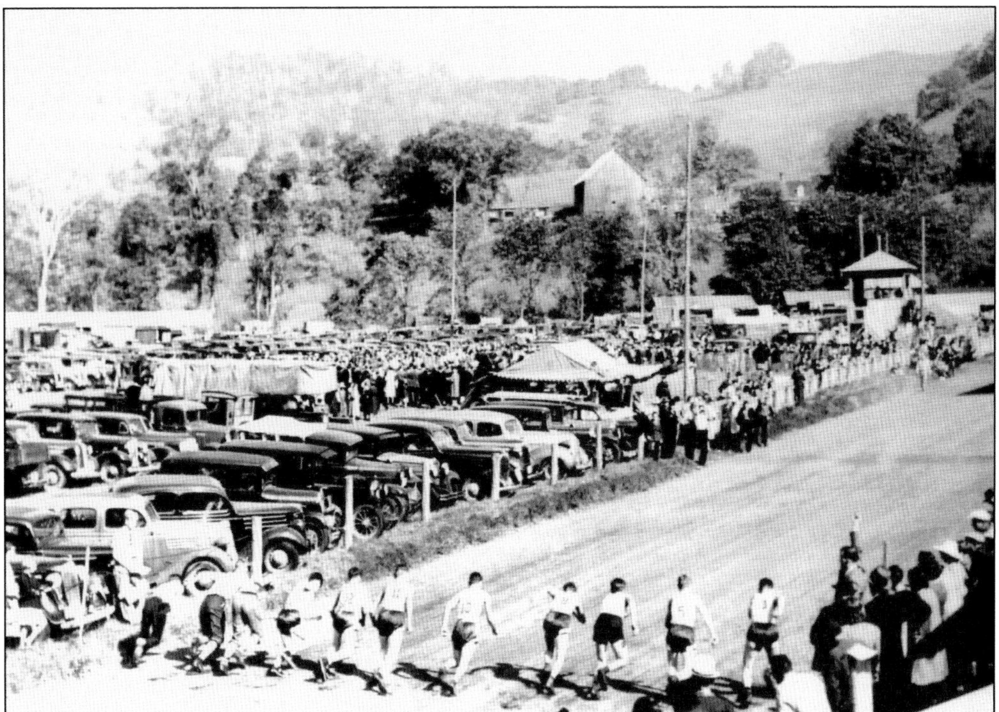

Prior to World War II, track and field events were very popular. Local high schools (South Royalton, Chelsea, Bethel, Randolph, and Rochester) entered teams. Spectators loved the intense rivalry.

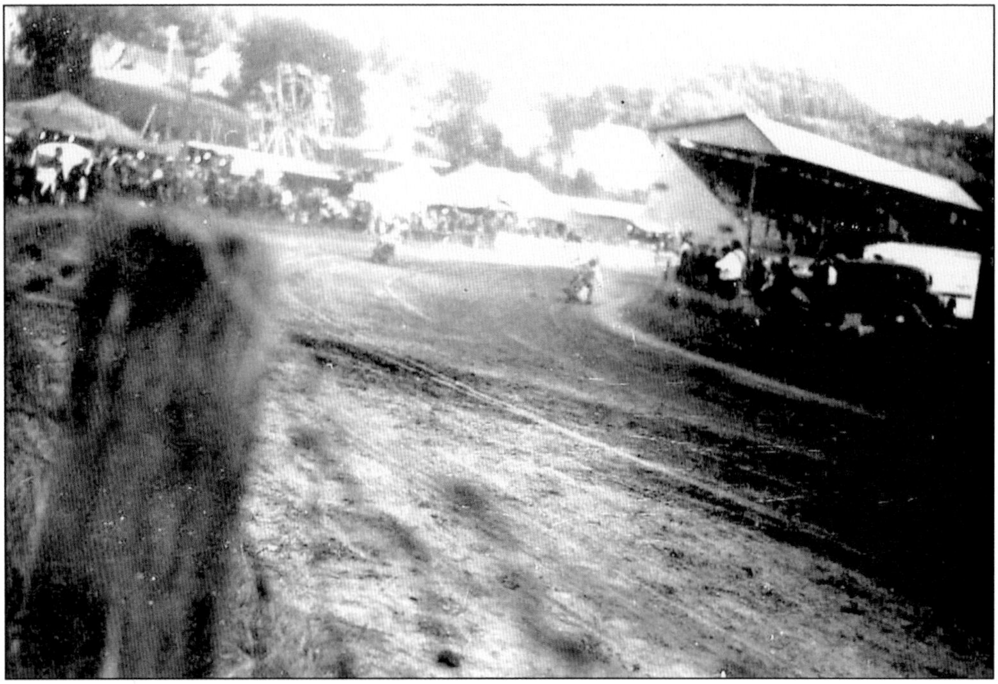

Motorcycle racing before World War II was a special attraction for just one year. The sharp turns on the dirt track were certainly an important reason why they did not return. There is just one known photograph of that entertainment.

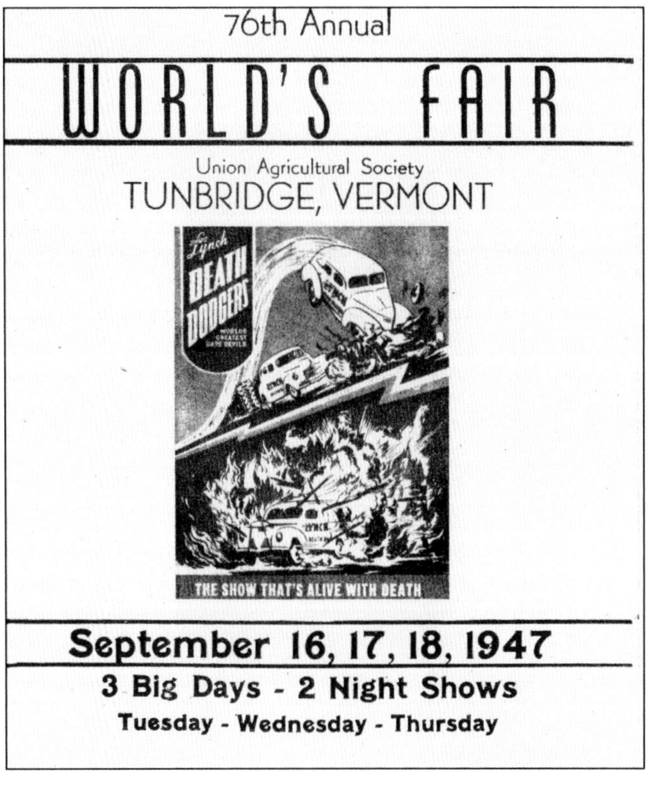

The 1947 program for the first time presented the "Death Dodgers" automobile thrill show, which sensationally was billed "The show that's alive with death." For the next 30 years, automobile thrill shows were a staple of the fair schedule.

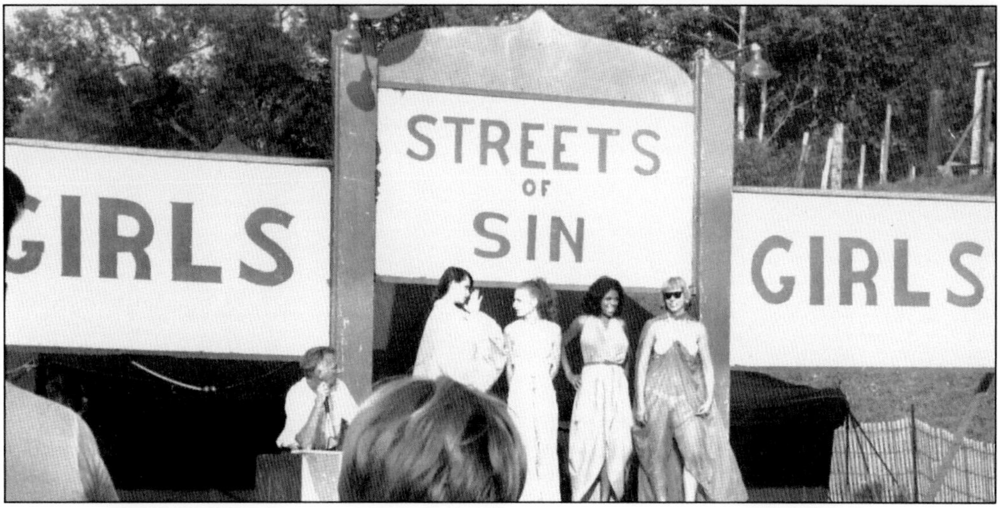

The post–World War II years brought many things to Vermont, some good and some not as good. One of the latter was the arrival of the "girlie shows" to the state's fairs. It seemed as though the "girls," with time, dropped even lower in caliber.

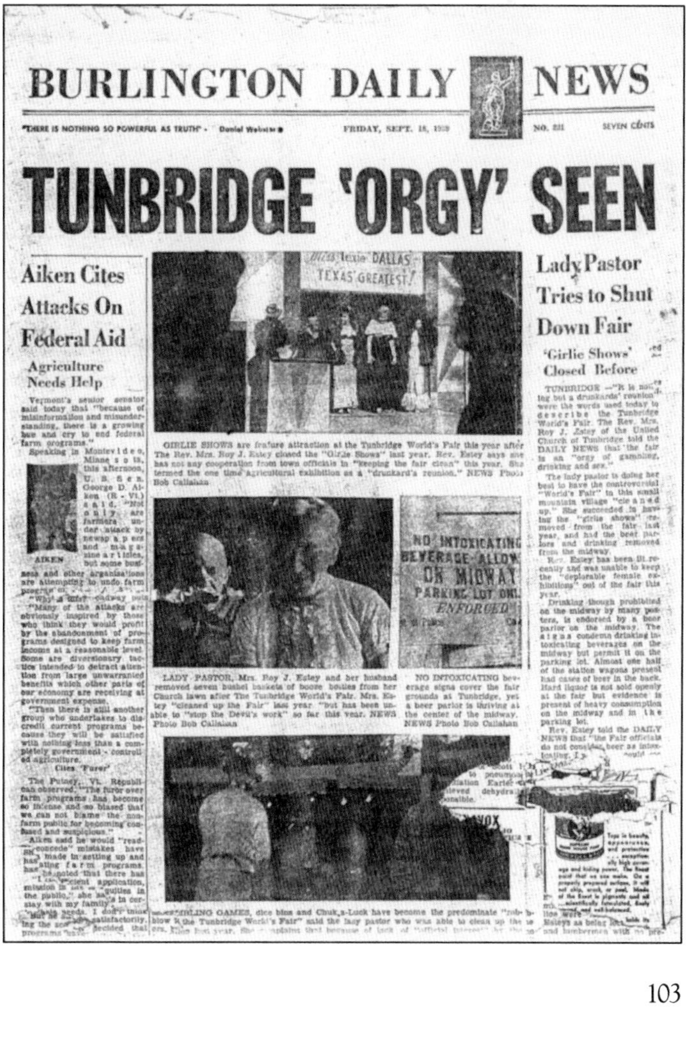

BURLINGTON DAILY NEWS

"THERE IS NOTHING SO POWERFUL AS TRUTH" - Daniel Webster FRIDAY, SEPT. 18, 1959 SEVEN CENTS

TUNBRIDGE 'ORGY' SEEN

Aiken Cites Attacks On Federal Aid

Agriculture Needs Help

Vermont's senior senator said today that "because of misinformation and misunderstanding, there is a growing hue and cry to end federal farm programs."

Speaking in Montevideo, Minnesota, this afternoon, U. S. Sen. George D. Aiken (R - Vt.) said, "Not only are farmers under attack by newspapers and magazine articles, but some business and other organizations are attempting to undo farm programs ...

AIKEN

"Many of the attacks are obviously inspired by those who think they would profit by the abandonment of programs designed to keep farm income at a reasonable level. Some are diversionary tactics intended to detract attention from large unwarranted benefits which other parts of our economy are receiving at government expense.

"Then there is still another group who undertakes to discredit current programs because they will be satisfied with nothing less than a completely government - controlled agriculture.

Cites 'Furor'

The Putney, Vt. Republican observed, "The furor over farm programs has become so intense and so biased that we can not blame the nonfarm public for becoming confused and suspicious."

Aiken said he would "readily concede" mistakes have been made in setting up and handling farm programs. He noted that there has ...

GIRLIE SHOWS are feature attraction at the Tunbridge World's Fair this year after The Rev. Mrs. Roy J. Estey closed the "Girlie Shows" last year. Rev. Estey says she has not any cooperation from town officials in "keeping the fair clean" this year. She termed the one time agricultural exhibition as a "drunkard's reunion." NEWS Photo Bob Callahan

LADY PASTOR, Mrs. Roy J. Estey and her husband removed seven bushel baskets of booze bottles from her Church lawn after The Tunbridge World's Fair. Mrs. Estey "cleaned up the Fair" last year "but has been unable to "stop the Devil's work" so far this year. NEWS Photo Bob Callahan

NO INTOXICATING beverage signs cover the fair grounds at Tunbridge, yet a beer parlor is thriving at the center of the midway. NEWS Photo Bob Callahan

Lady Pastor Tries to Shut Down Fair

'Girlie Shows' Closed Before

TUNBRIDGE —"It is nothing but a drunkards' reunion" were the words used today to describe the Tunbridge World's Fair. The Rev. Mrs. Roy J. Estey of the United Church of Tunbridge told the DAILY NEWS that "the fair is an "orgy of gambling, drinking and sex."

The lady pastor is doing her best to have the controversial "World's Fair" in this small mountain village "cleaned up." She succeeded in having the "girlie shows" removed from the fair last year, and had the beer parlors and drinking removed from the midway.

Rev. Estey has been ill recently and was unable to keep the "deplorable female exhibitions" out of the fair this year.

Drinking though prohibited on the midway by many posters, is endorsed by a beer parlor on the midway. The signs condemn drinking intoxicating beverages on the midway but permit it on the parking lot. Almost one half of the station wagons present had cases of beer in the back. Hard liquor is not sold openly at the fair but evidence is present of heavy consumption on the midway and in the parking lot.

Rev. Estey told the DAILY NEWS that "the Fair officials do not consider beer as intoxicating. I ...

... GAMES, dice bins and Chuk-a-Luck have become the predominate ... the Tunbridge World's Fair" said the lady pastor ...

The *Burlington Daily News* (now defunct) issue of September 18, 1959, devoted much of the front page to Rev. Katherine Estey's effort to close down the fair because of drunkenness and the girlie shows. She was not successful and they continued for another 20 years. Finally in 1979 the fair directors made some major changes including the discontinuation of the girlies. Now the fair is one of the best agricultural and family events in the state.

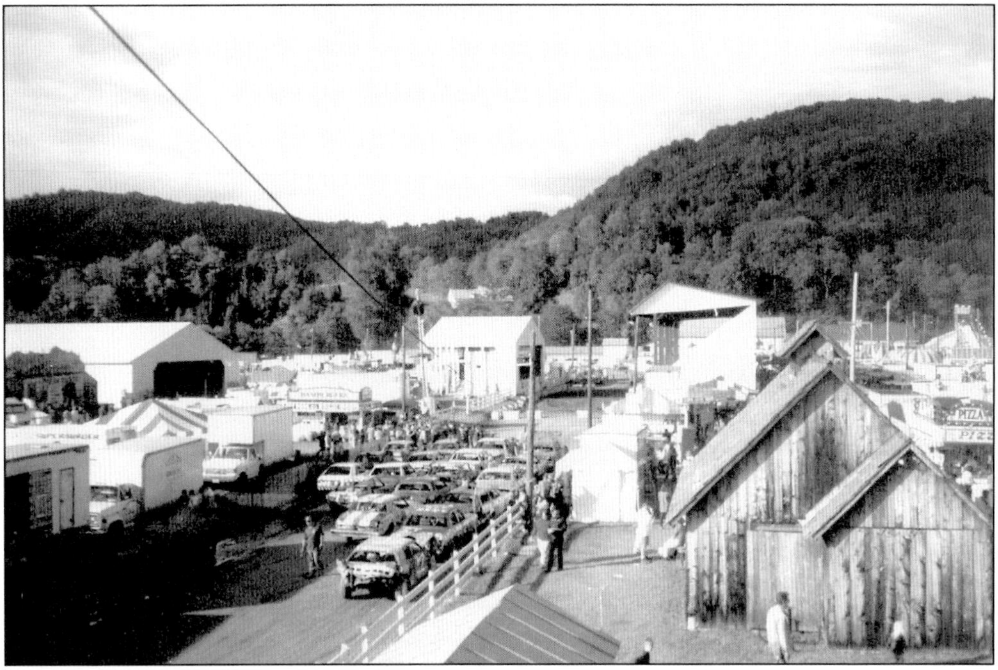

A demolition derby first came to the fair in 1979. It was always scheduled on Sunday afternoon. Here the cars are lining up for a heat.

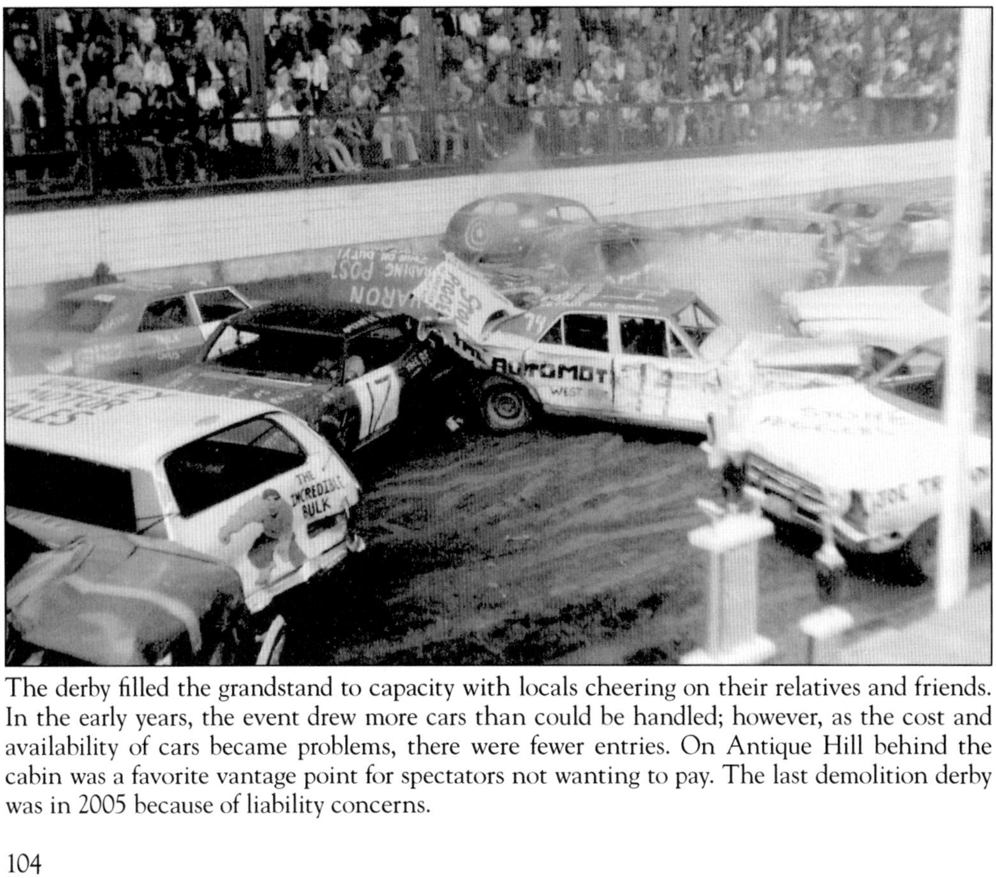

The derby filled the grandstand to capacity with locals cheering on their relatives and friends. In the early years, the event drew more cars than could be handled; however, as the cost and availability of cars became problems, there were fewer entries. On Antique Hill behind the cabin was a favorite vantage point for spectators not wanting to pay. The last demolition derby was in 2005 because of liability concerns.

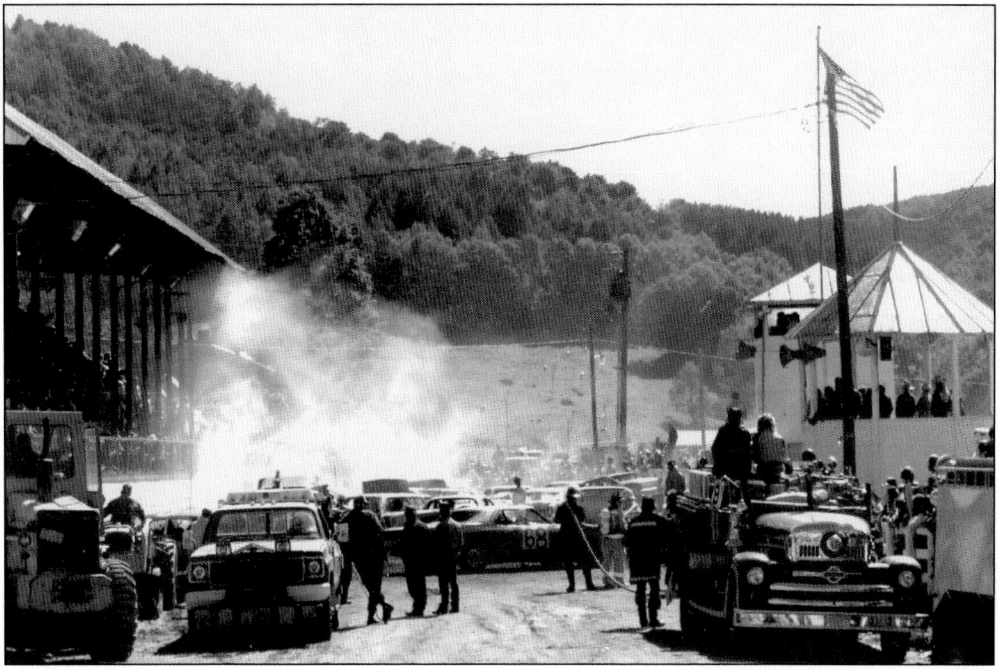

The Tunbridge Volunteer Fire Department was standing by in case there were engine fires. The rising smoke in the above photograph shows such a fire. (Courtesy of Arlene Stockwell.)

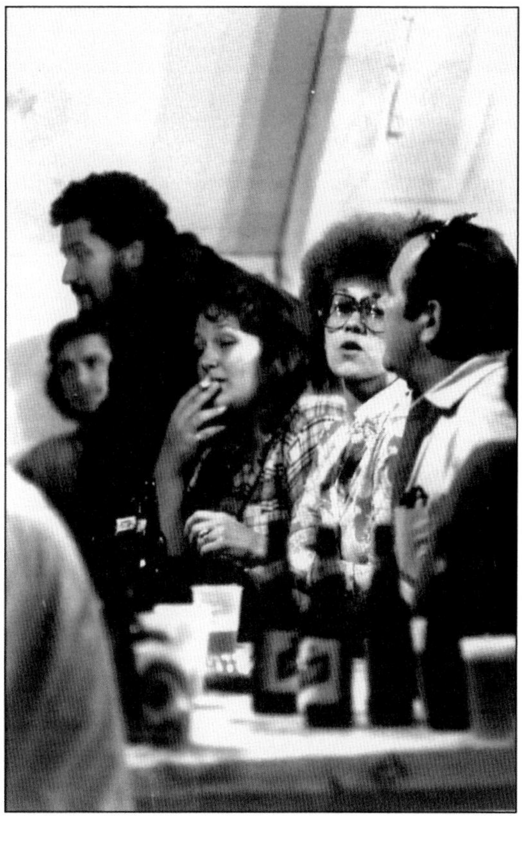

In 1952, the directors voted to sell beer at the fair. The beer hall was opened under the grandstand. While it is not a charming place (chicken wire over the windows), people wax nostalgic about it. Years ago, beer was still served in bottles and smoking was allowed.

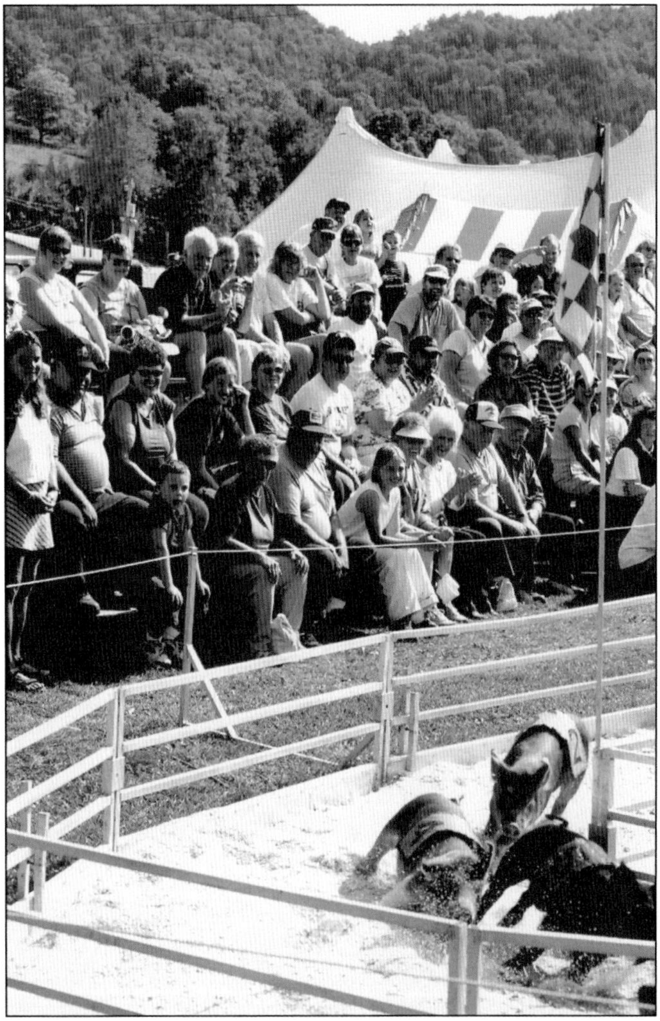

The pig races are heavily attended. It is fun to watch the pigs but the barker is the real show. At the end of the race they all receive their Oreo cookies.

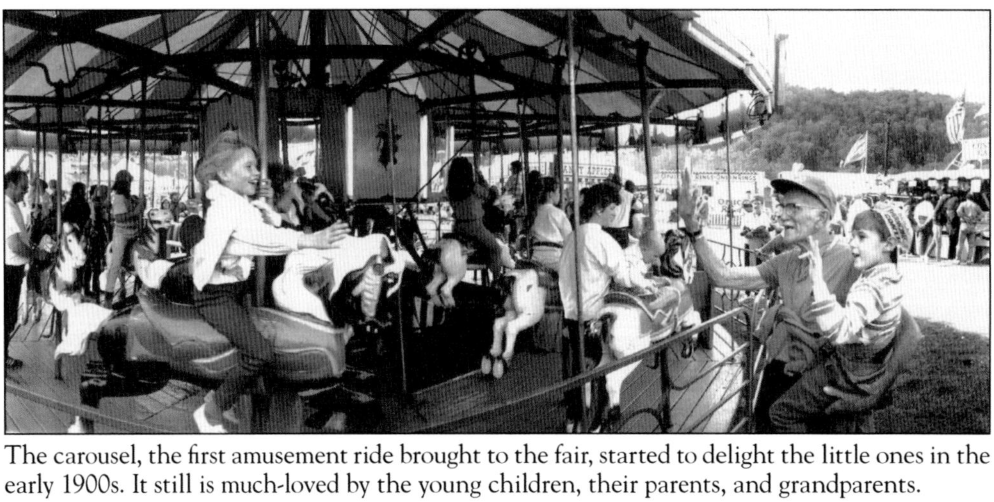

The carousel, the first amusement ride brought to the fair, started to delight the little ones in the early 1900s. It still is much-loved by the young children, their parents, and grandparents.

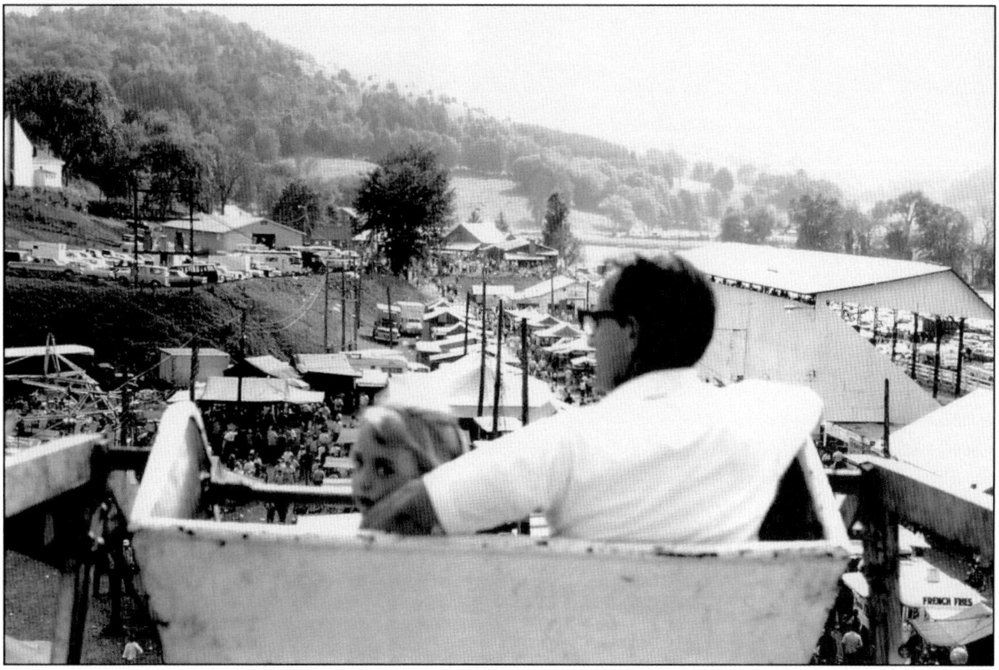

This 1969 shot may have been taken on a Saturday since the midway is full. When the Ferris wheel paused the passengers had a terrific view of the fair. Concessions still worked out of tents, but within 20 years, they had all become self-contained.

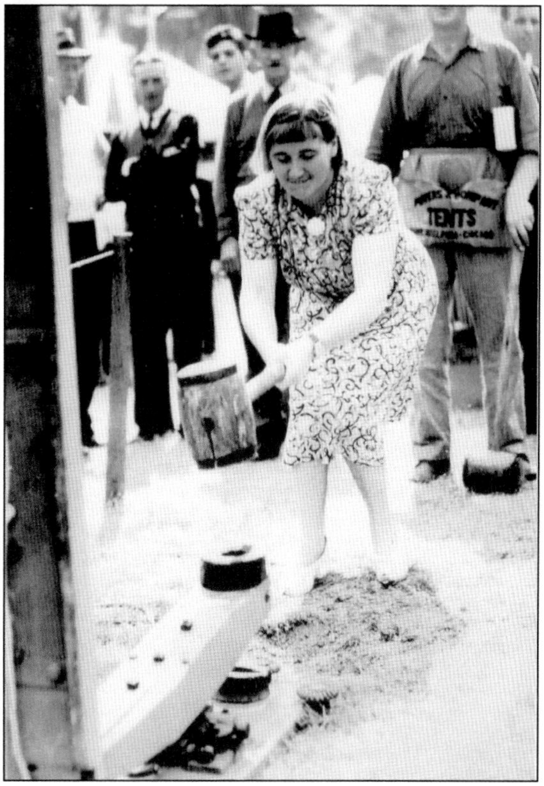

An unidentified woman is attempting to win a prize by ringing the bell. Folks have long surmised that the operator tampered with the apparatus so that "weak" females succeeded and "macho men" failed.

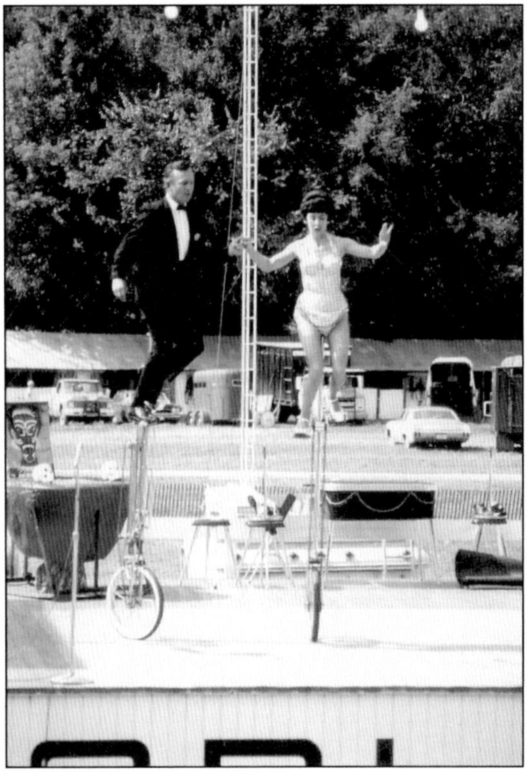

Vaudeville acts used to be alternated with sulky races on the program. Several different acts were yearly part of the entertainment.

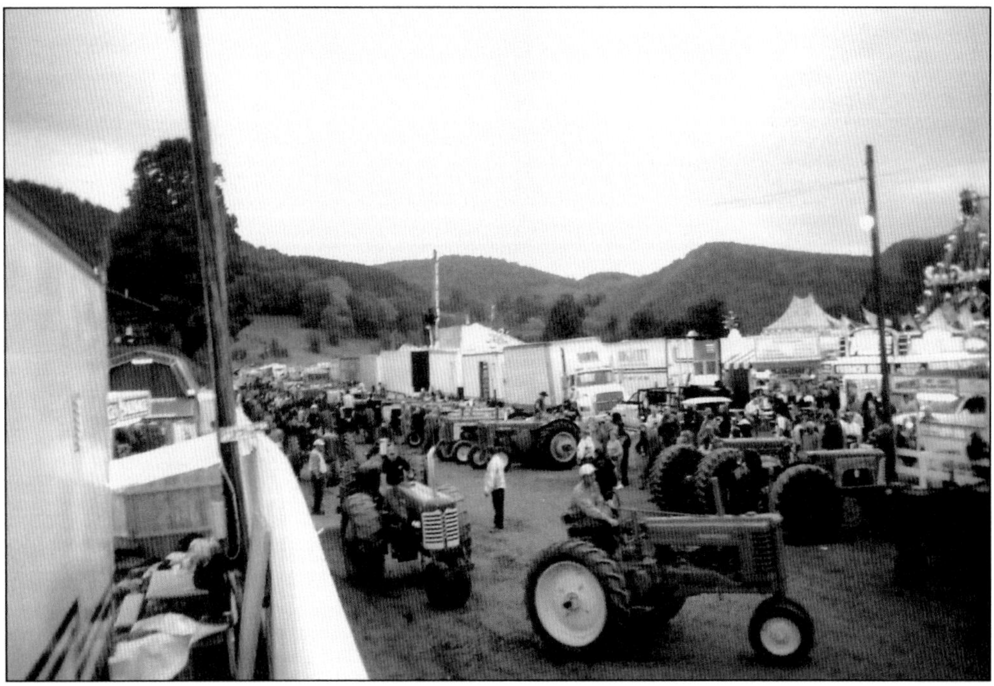

An antique tractor pull has become heavily attended on Friday nights at the fair. Dozens of tractors of all descriptions participate. The 1,200 seat grandstand is full even though the competition often runs for several hours.

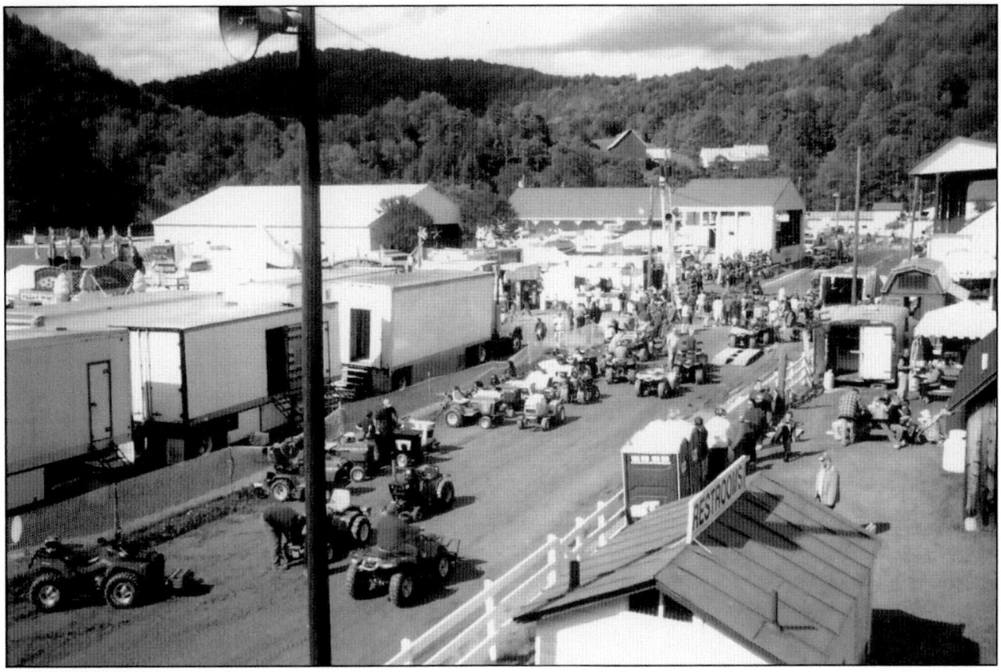

ATV pulling and lawn tractor pulling replaced the demolition derby in 2006 and became an instant success.

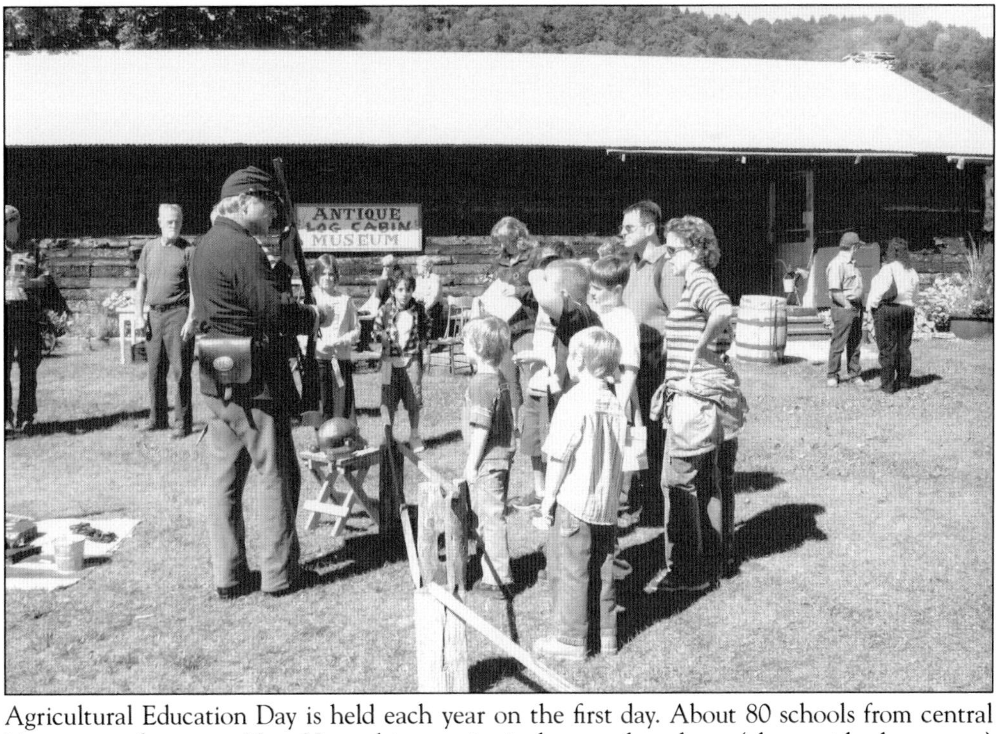

Agricultural Education Day is held each year on the first day. About 80 schools from central Vermont and western New Hampshire are invited to send students (along with chaperones) free of charge to the fair. A day-long series of agriculture-related happenings are put on for the children. A Civil War reenactor, above, explains his musket to a group.

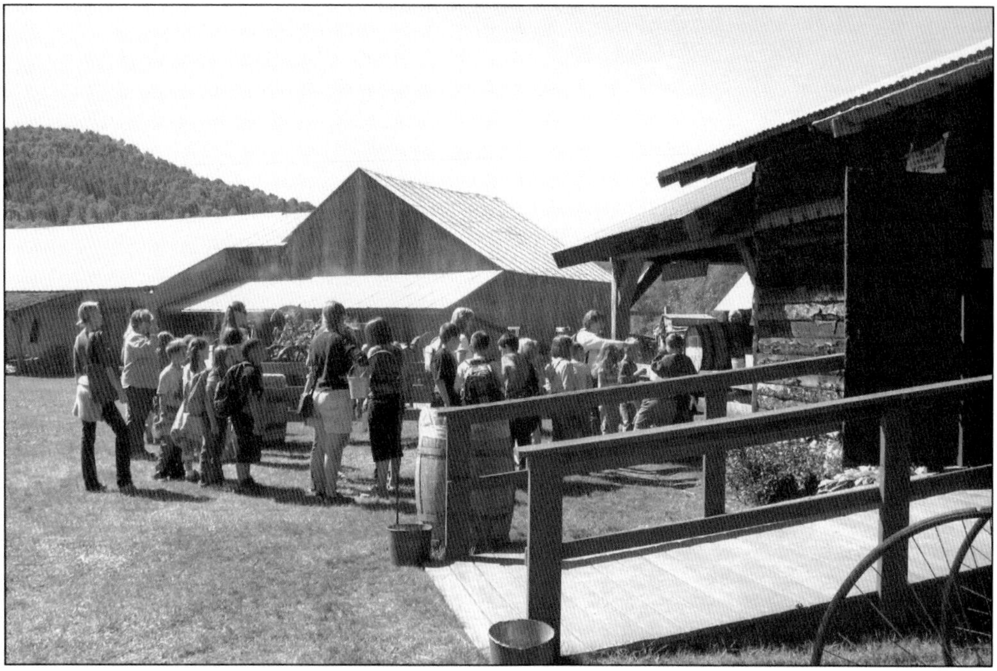

Several thousand children come from public and private schools, as well as home-schooled children. They all enjoy getting a cup of freshly pressed cider.

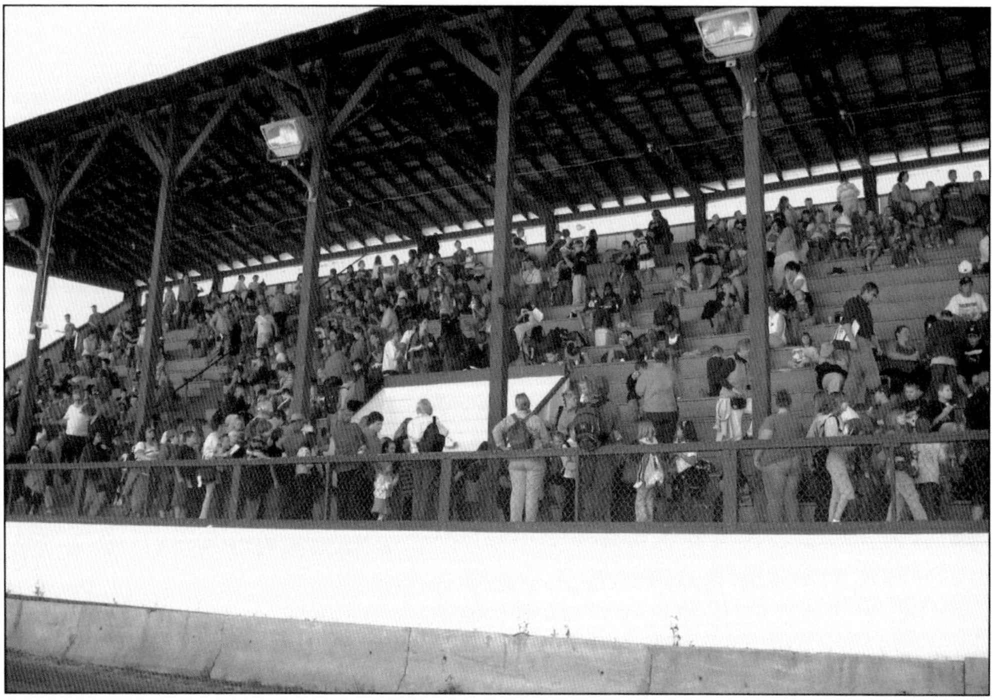

The children, teachers, and chaperones are invited to eat their lunch in the grandstand. Free milk is provided for everyone. Agricultural Education Day has proved to be one of the most innovative and successful days the fair has sponsored. Many teachers remark that it is their favorite day of the school year.

Five

PEOPLE

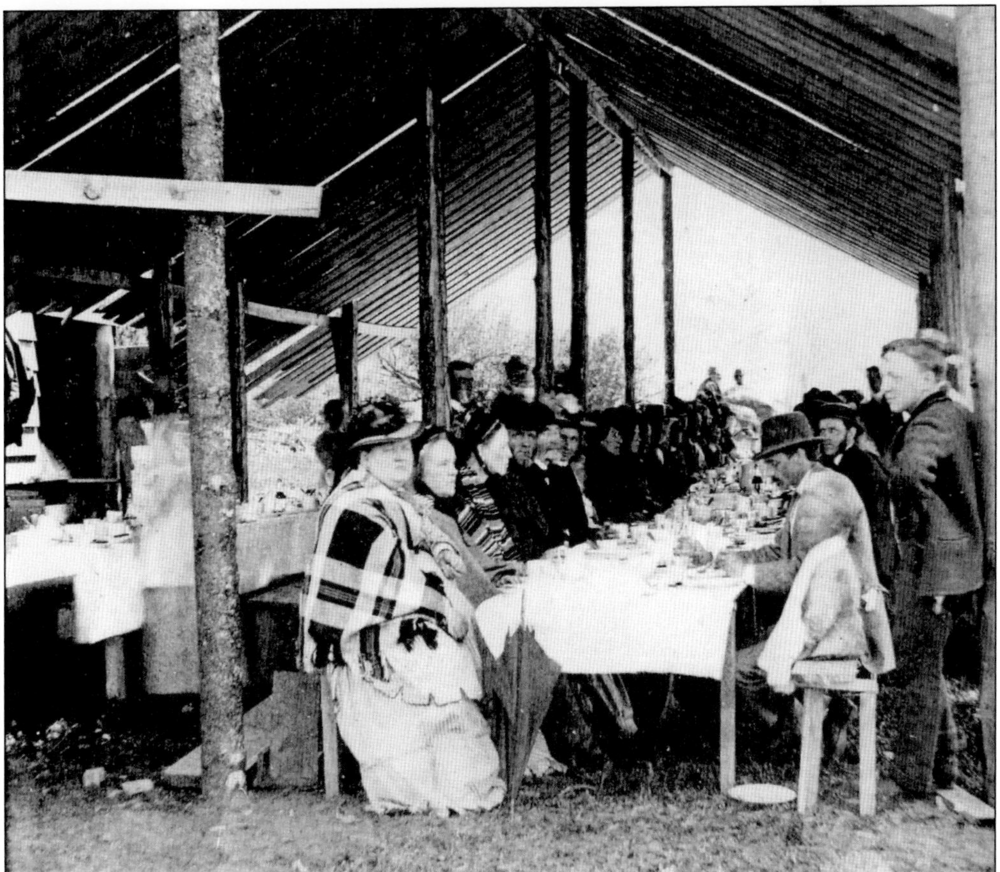

The only known photograph (stereograph) taken at the original fairgrounds in North Tunbridge shows an eating shed, which was basically a pole barn with a tin roof. It was very likely the only building on Lougee's Field.

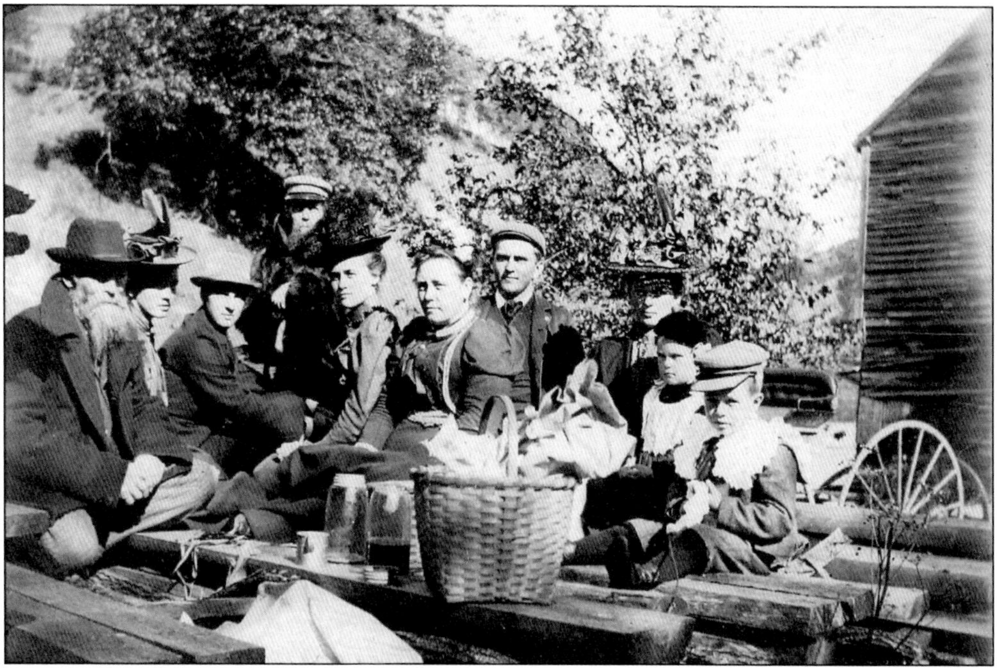

One of the real pleasures of the early fairs was that it brought together family members. Three generations of the Dexter Larkin clan are enjoying the occasion, including patriarch Dexter, extreme left, his wife Ellen behind the basket, and Alice Durkee (Farnham), second from right. Notice that the man standing in back appears to be wearing a bear skin coat.

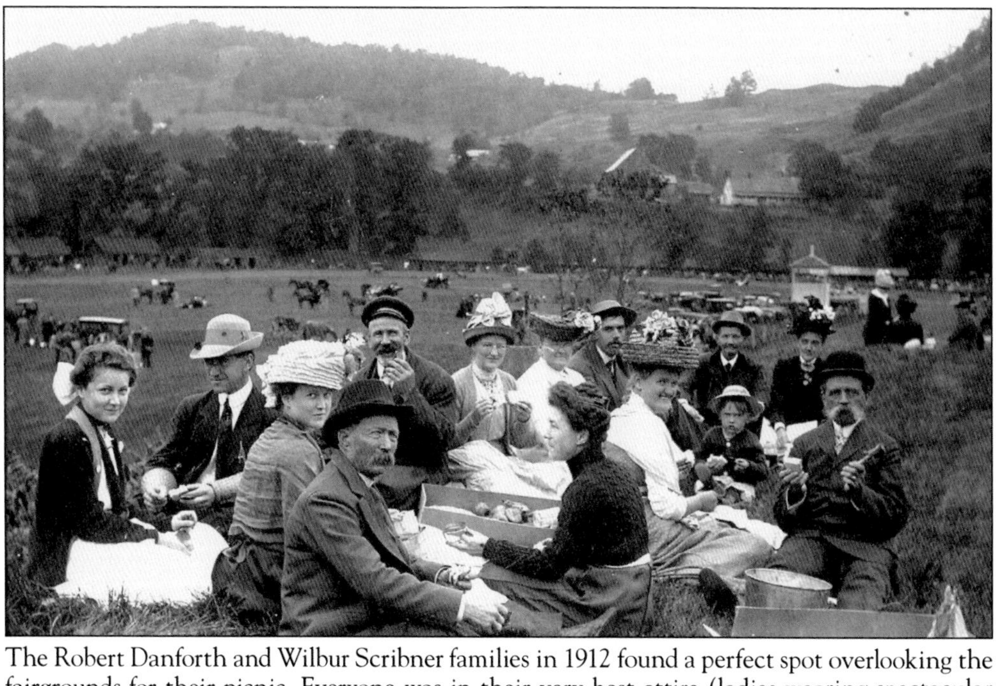

The Robert Danforth and Wilbur Scribner families in 1912 found a perfect spot overlooking the fairgrounds for their picnic. Everyone was in their very best attire (ladies wearing spectacular hats) whenever they attended functions. They probably had just one good suit of clothes. (Courtesy of Bill and Marie Danforth.)

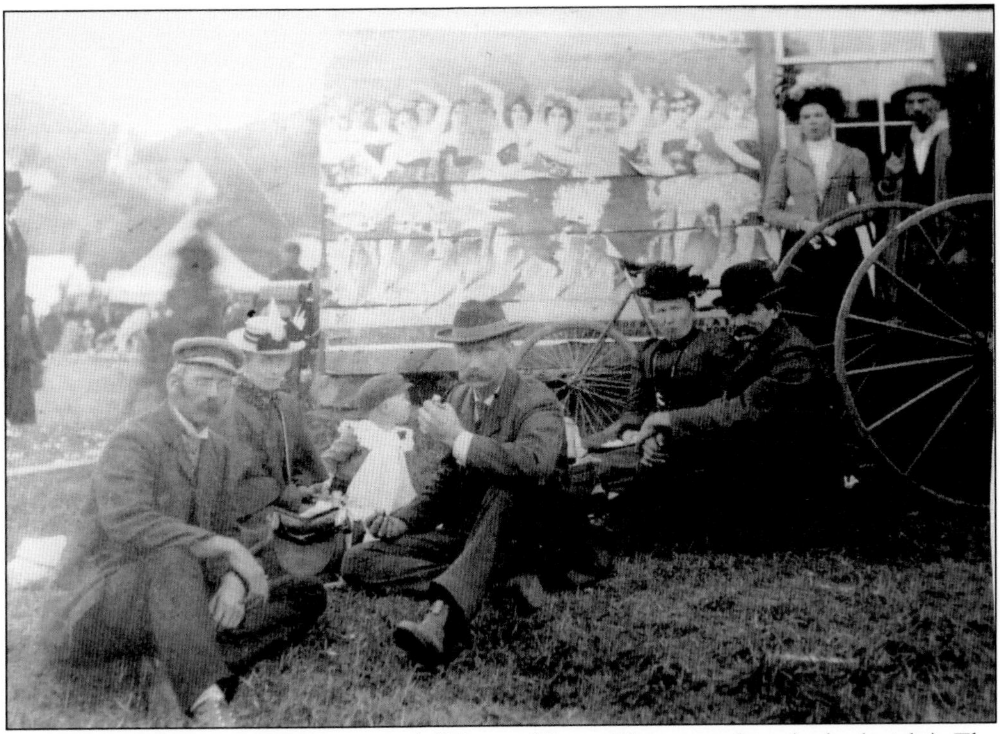

The Glines family group shares a meal (John and Lucy Glines stand in the back right). The mystery of this picture concerns the poster of dancing women, which perhaps advertised a musical review. Most families were large, hard working, and not moneyed, so bringing extensive picnic baskets was a time-honored custom. Notice the little boy is dressed in a ruffled pinafore.

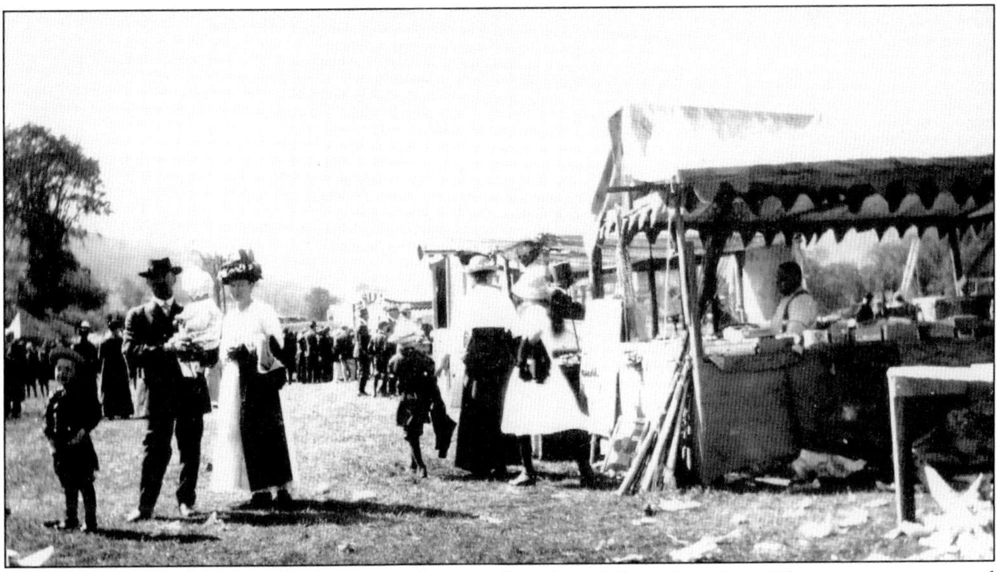

Each family member, on the left, is garbed in their best bib and tucker. The concession stand signs are not legible, but oyster bars, tintype booths, and fortune tellers were all popular.

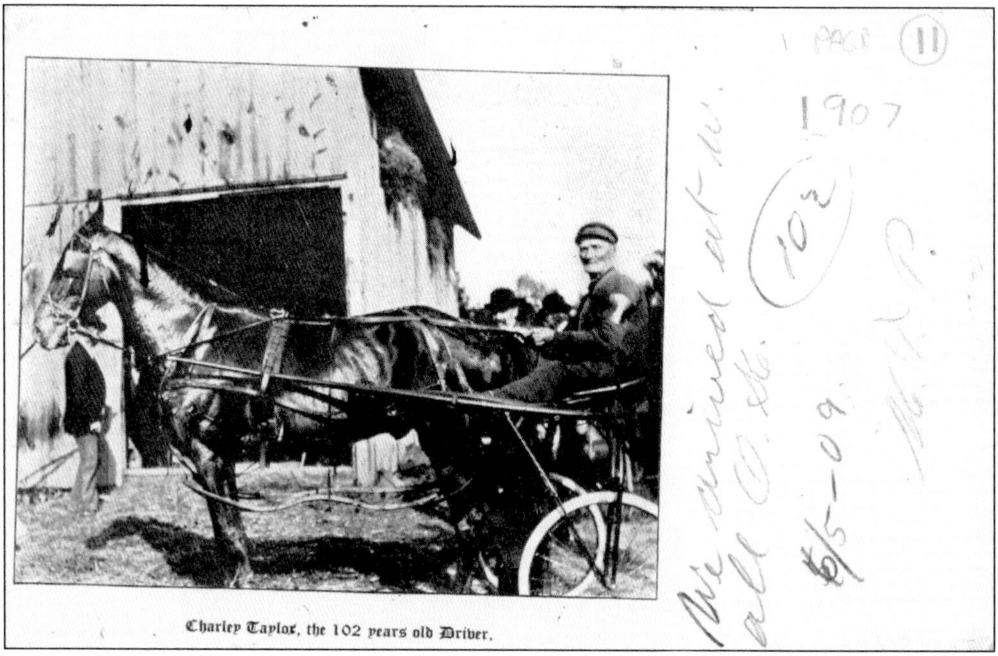

An old postcard honored "Uncle Charley" Taylor of White River Junction, who most likely was the oldest driver who raced on the Tunbridge track. At the age of 92 in 1898, he entered the free-for-all with his horse Robert B. and won two heats. He made a track record of 2:18.4 in 1899. When Taylor was 102 he drove exhibition miles at both Tunbridge and White River Junction fairs.

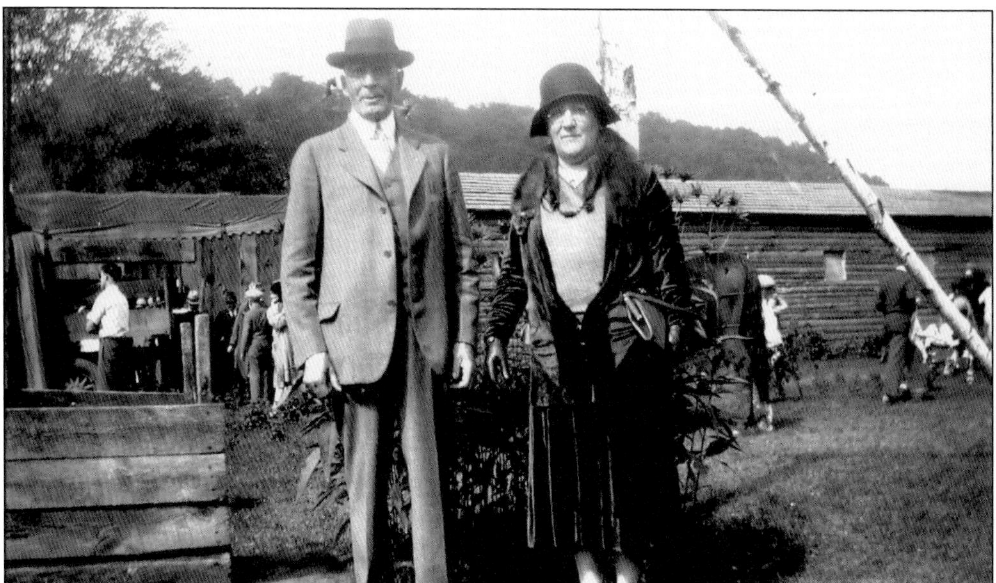

A couple on Antique Hill in the 1930s came to the fair in their Sunday-go-to-meeting ensembles. Her stylish velvet suit is completed by a mink piece, cloche hat, and kid gloves. He (pipe in mouth) was in a fine three-piece suit.

The South Tunbridge Methodist Episcopal Church, starting in 1931 and continuing for some 20 years (excluding World War II), operated a baked bean booth under the grandstand. Quickly it became "one of the best places to eat," evidenced by the line, left, and the satisfied customer, right.

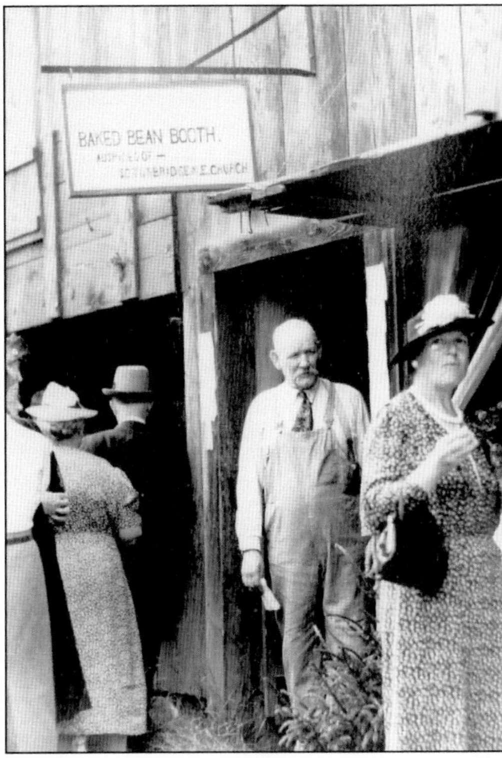

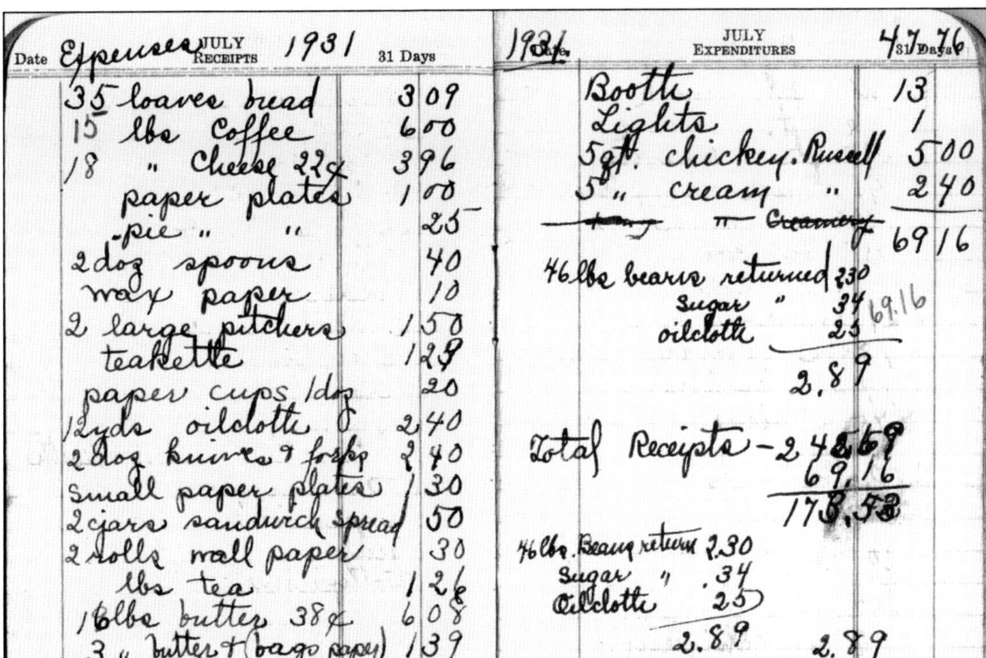

Date Expenses JULY RECEIPTS 1931	31 Days
35 loaves bread	3 09
15 lbs Coffee	6 00
18 " Cheese 22¢	3 96
paper plates	1 00
pie " "	25
2 doz spoons	40
wax paper	10
2 large pitchers	1 50
teakettle	1 29
paper cups 1 doz	20
12 yds oilcloth	2 40
2 doz knives & forks	2 40
small paper plates	1 30
2 jars sandwich spread	50
2 rolls wall paper	30
lbs tea	1 26
16 lbs butter 38¢	6 08
3 " butter & (bags paper)	1 39

1931 JULY EXPENDITURES	47.76 31 Days
Booth	13
Lights	1
5 qt. chicken. Russell	5 00
5 " cream "	2 40
	69 16
46 lbs beans returned	2.30
sugar "	34
oilcloth	25
	2.89
Total Receipts –	24[illegible].69
	69.16
	17[illegible].53
46 lbs. Beans return	2.30
sugar "	.34
oilcloth "	.25
	2.89
	2.89

Marion Howe, wife of James K. Howe, kept financial records of the church's booth. The first year they sold 54 pounds of baked beans and used 15 pounds of coffee. The expenses totaled $69.16 and the profit was $176.42, which was divided between the Bible class and the 3M's (the youth group). (Courtesy of David Howe.)

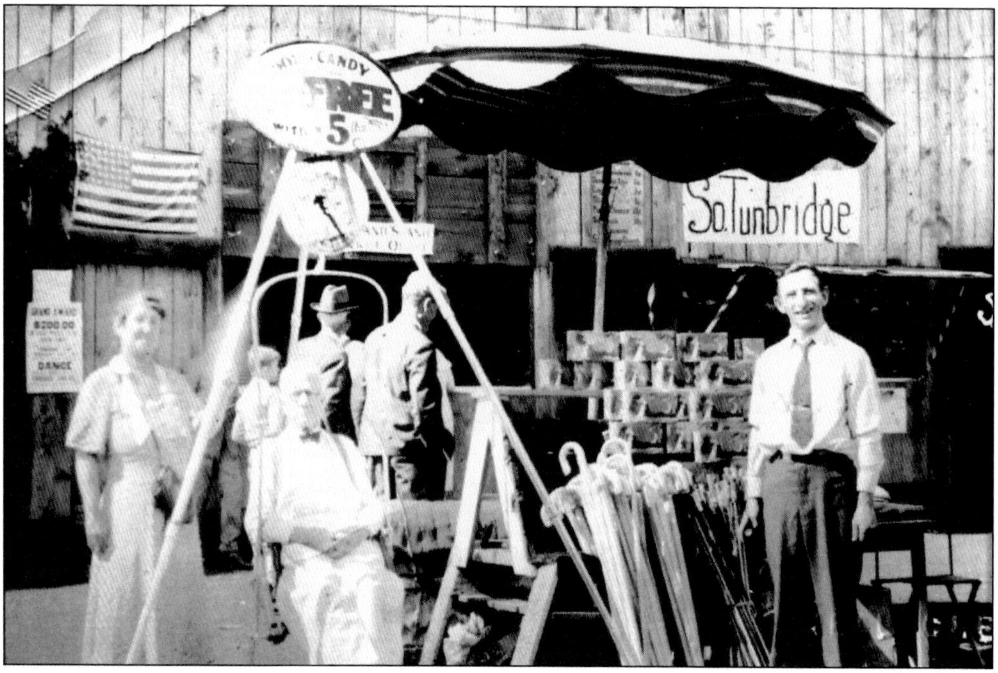

The Rosenbaum family from New York City had a "guess your weight or age" concession stand for a long time. A 1934 image shows all three of them at their prime location, which was in back of the grandstand. The Rosenbaums boarded at the Tunbridge Village home of the author's Durkee grandparents until World War II.

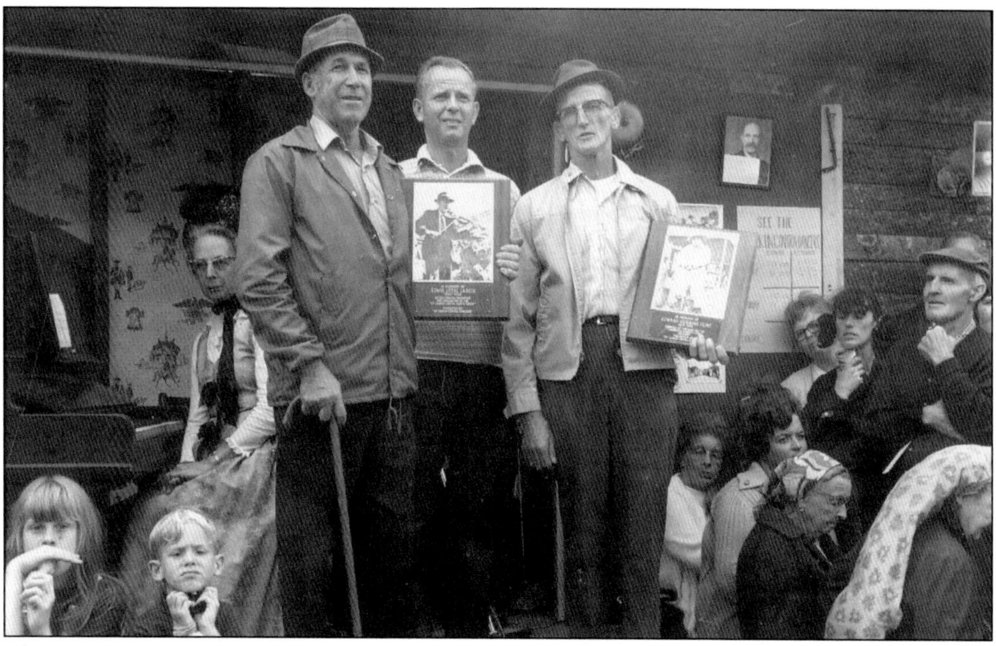

The 1973 antique committee, from left to right, John Howe, David Cilley, and Henry Dodge, displayed the plaques honoring Ed Flint and Ed Larkin, which were presented by the Ed Larkin Contra Dancers. The two men contributed greatly in making Antique Hill the success it became. Gertrude Larkin Roberts, Larkin's daughter, is at the piano.

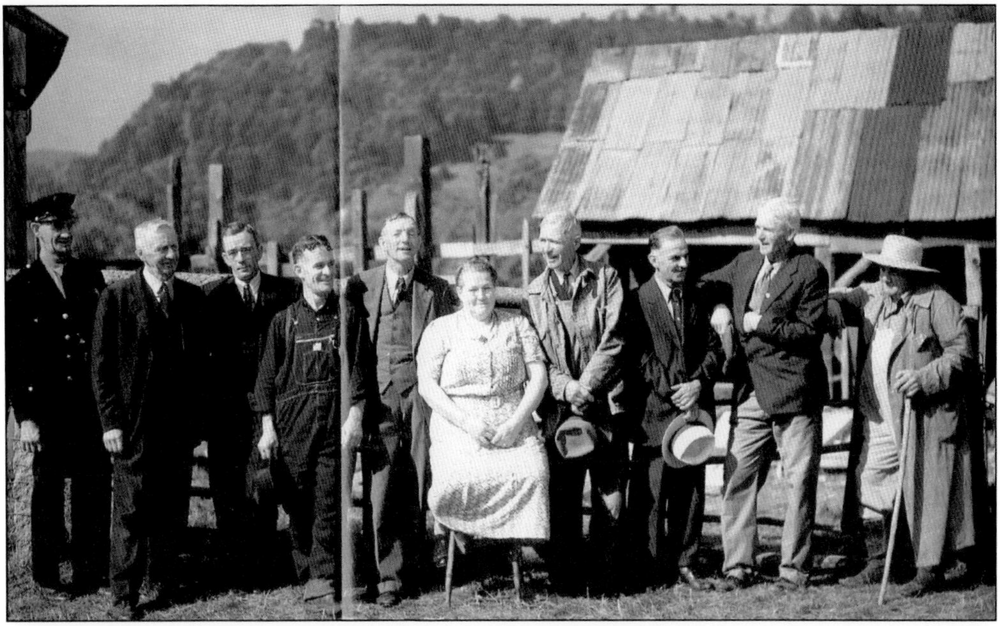

A wonderful portrait of the board of directors of the Union Agricultural Society taken in 1941 by Warren Dexter shows, from left to right, Oliver Burgess, Orange County sheriff; Frank Austin, vice president; Phil Farnham; Herman Ashline; Arthur Ellis; Marion Howe, treasurer (not a board member); James K. Howe; Dean Camp; Nathan Dodge, president; and Ed Flint, secretary. This photograph appeared in the 50th anniversary issue of *Vermont Life* magazine, autumn 1996. (Courtesy of Robert Dexter.)

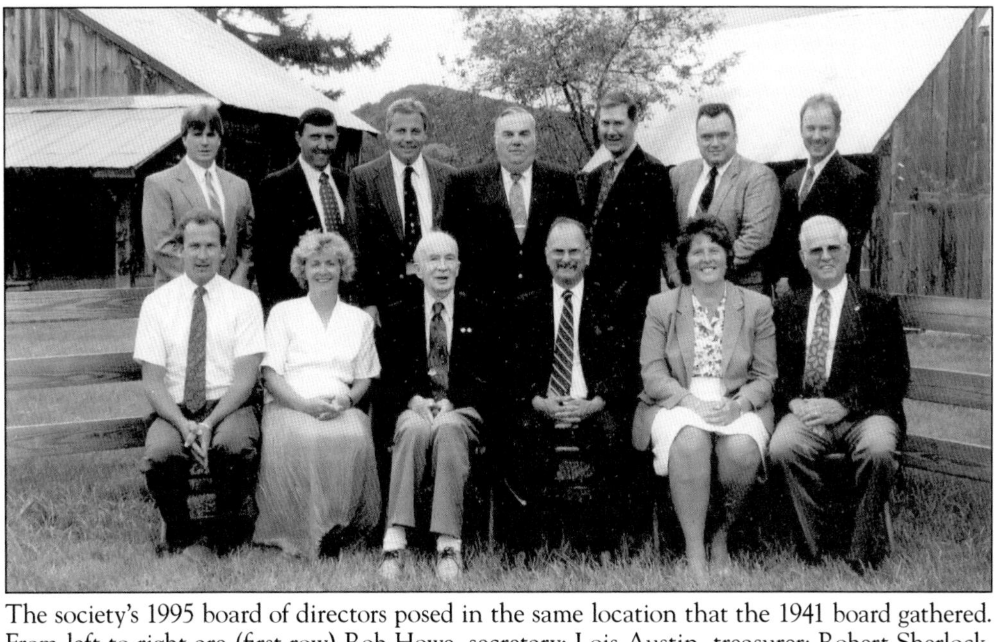

The society's 1995 board of directors posed in the same location that the 1941 board gathered. From left to right are (first row) Rob Howe, secretary; Lois Austin, treasurer; Robert Sherlock, director emeritus and former secretary; Euclid Farnham, president; Betty Rogers; and David Cilley; (second row) Dennis Cilley; Gordon Barnaby, vice president; Wayne Ladd; Dana Goodnow; David Howe; Ben Tucker; and John Durkee.

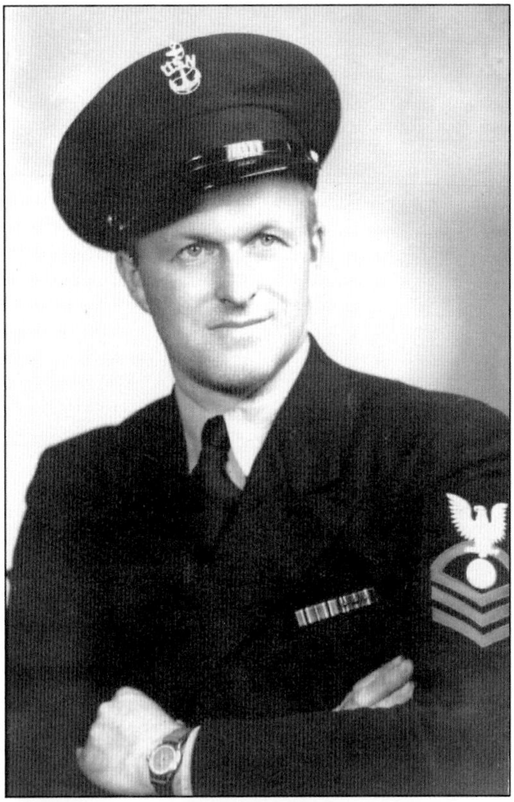

The Austin family historically has had the fair in its blood. Nathaniel Austin, who was involved from the fair's inception in 1867, served as president 1902 to 1916. Frank followed in his father's footsteps as president 1945 to 1962. He was also the superintendent of Floral Hall for some 40 years. Nellie Austin, Frank's wife, was Floral Hall secretary for most of that time. Wayland Austin (seen during World War II), son of Frank and Nellie, was director and superintendent of police and concessions 1962 to 1978. Fourth-generation Lois Austin has been director and treasurer since 1994.

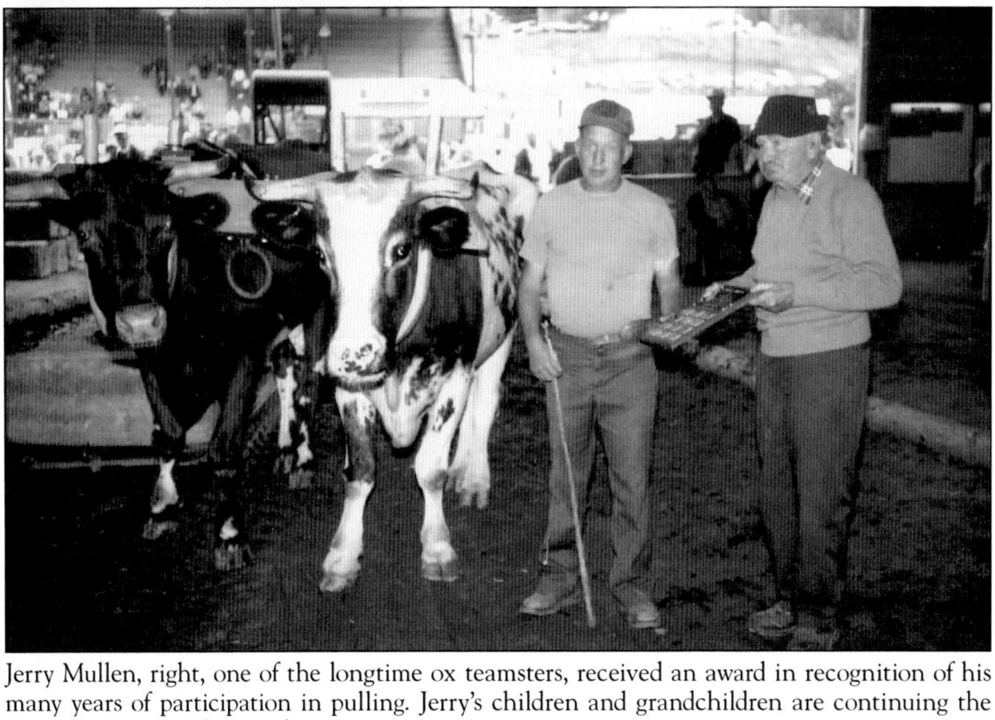

Jerry Mullen, right, one of the longtime ox teamsters, received an award in recognition of his many years of participation in pulling. Jerry's children and grandchildren are continuing the family tradition of fair involvement.

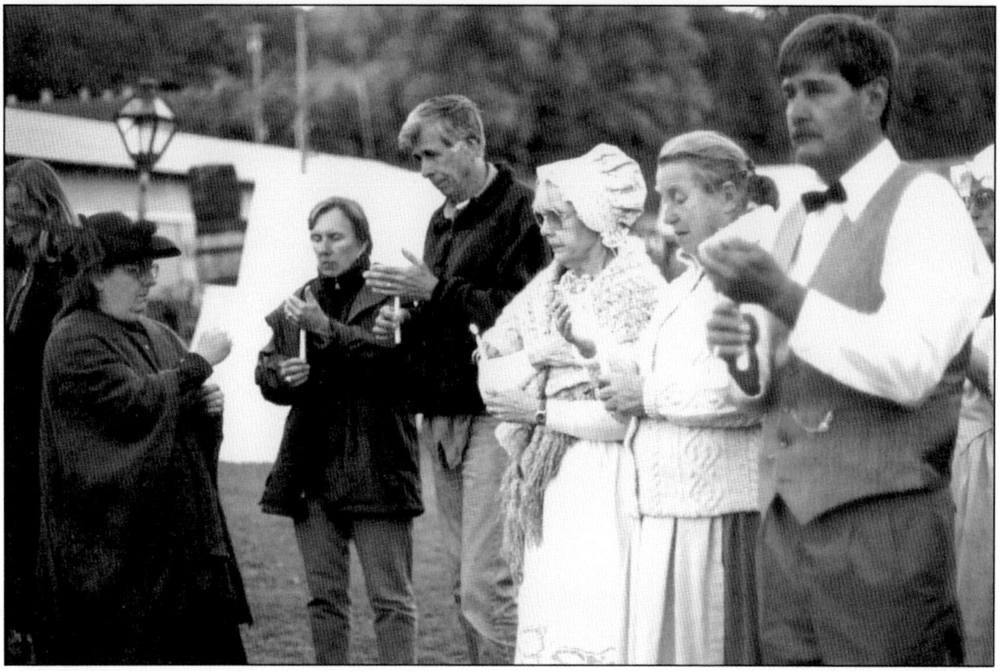

The World Trade Center was attacked September 11, 2001, two days before the scheduled fair. Candlelight vigils were arranged for dusk the first two evenings of the fair. Detractors declared that it could not be done; however, those few silent moments turned out to be the most moving times that have occurred on the grounds. During the hushed 10 minutes on Friday night, only the occasional bawling of a cow could be heard.

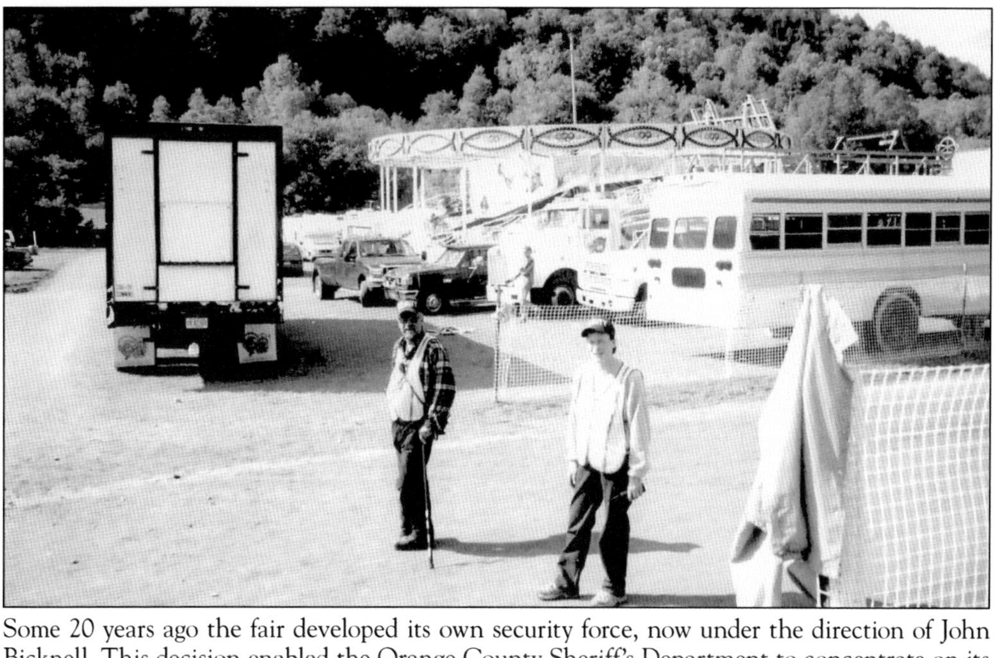

Some 20 years ago the fair developed its own security force, now under the direction of John Bicknell. This decision enabled the Orange County Sheriff's Department to concentrate on its police work. Security responsibilities include directing traffic, monitoring the grandstand, and guarding the racetrack.

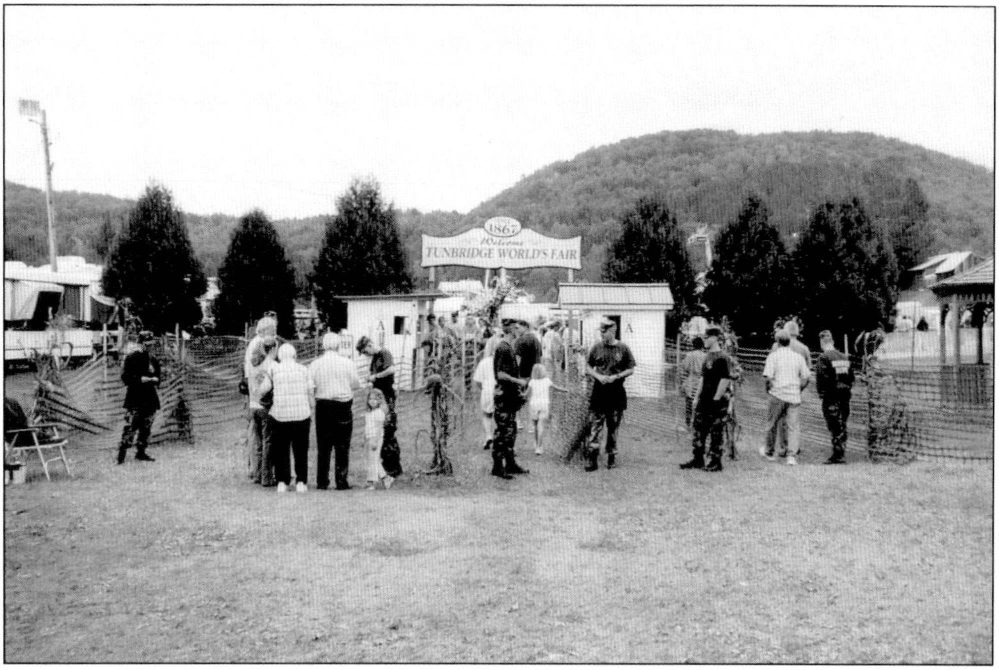

The U.S. Navy and Marine cadets at Norwich University in 1994 were contracted to sell tickets, and later to park cars during the fair. This has been a "win-win" situation for both parties. The fair needed additional staff and the Norwich cadets needed money for their charitable work. Fairgoers are appreciative of the efficient and courteous demeanor shown by these fine young men and women.

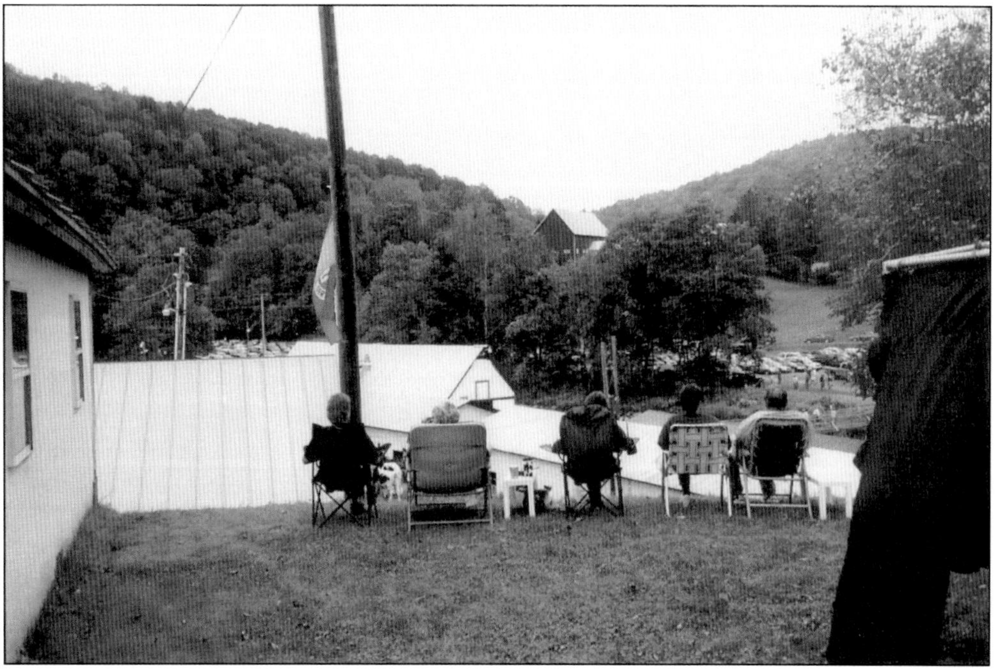

The First Branch Ambulance and Rescue organization provides excellent medical services. During breaks they enjoy the most private spot on the fairgrounds.

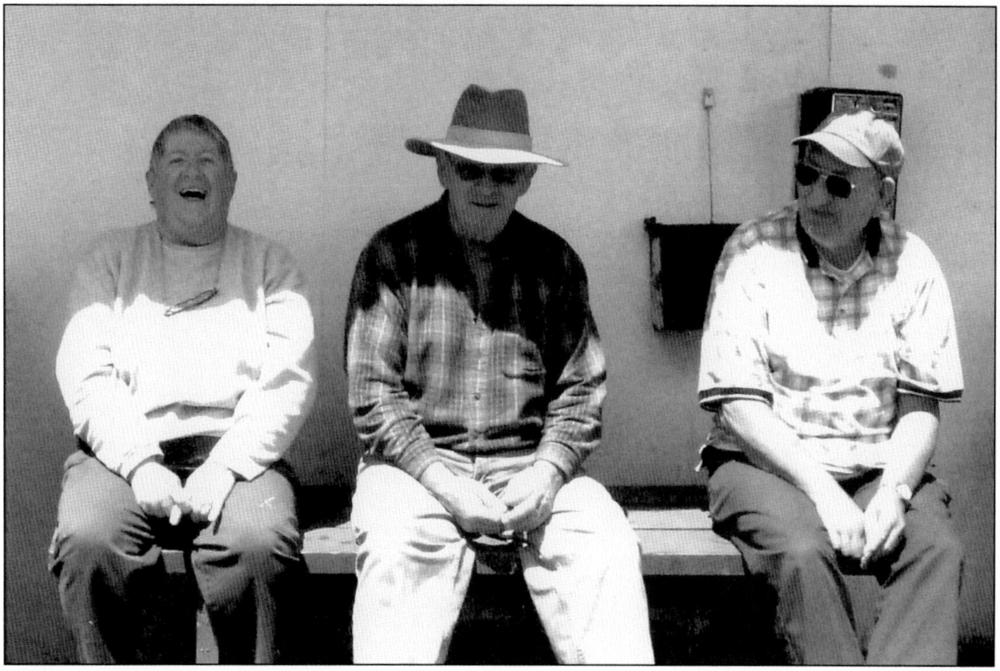

Nellie Goewey, left, grounds manager Cliff Goewey, center, and Frank Ricker take a break from their busy schedule for a hearty laugh. Cliff Goewey and his crew do a great job maintaining the grounds, preparing for numerous summer events and keeping the area a beautiful place. Kevin Barnaby is not pictured. (Courtesy of Tess Taylor.)

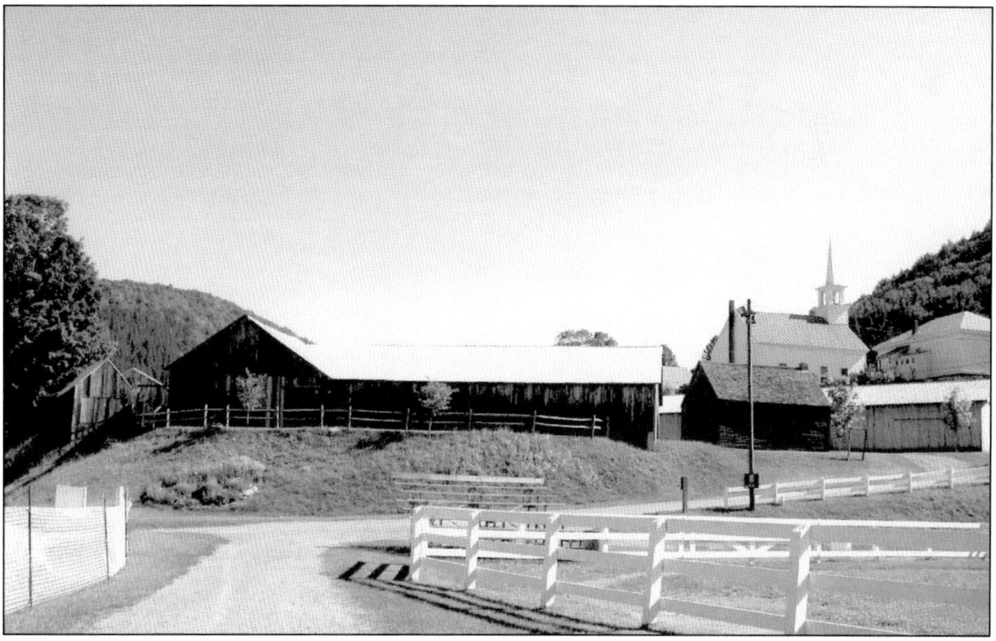

A dedicated group of retired staff have helped care for the property, including Richard "Joe" Rogers, Edward "Skee" Barnaby, and Bob Gray. Now that the fair has a full summer schedule, it is vital to keep the grounds in good repair and attractive. The first view of Antique Hill upon entering the fair is a testimonial to their work.

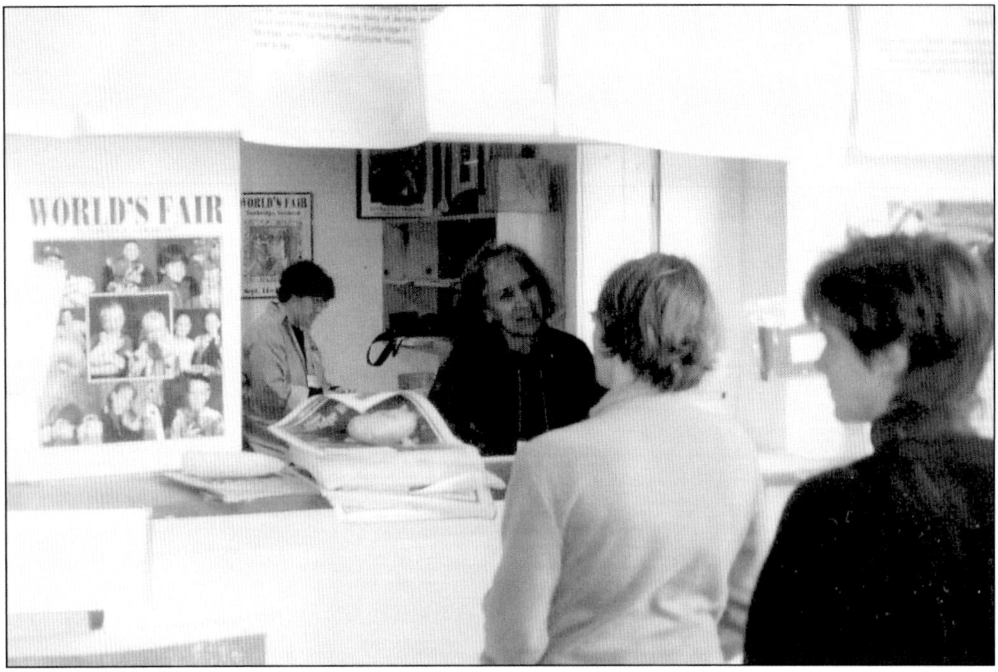

Tracy Amell, left, and Liz Guth, right, expertly run the registration office located at Dodge-Gilman. These two ladies process all entries leading up to the fair. They enjoy dispersing the premium checks the last day.

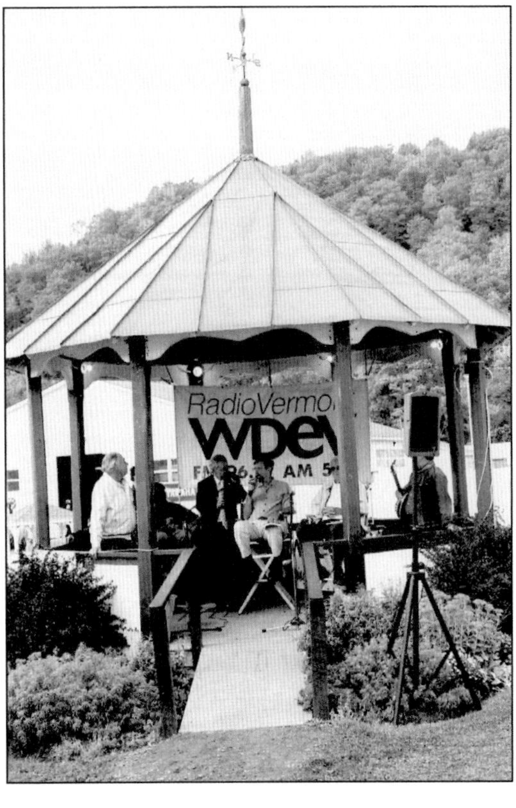

WDEV Radio in recent years has named Friday as Chautauqua Day. They set up in the old bandstand, now called the gazebo. Conversations with fairgoers and Vermont experts can range from agriculture in the Northeast to regional music.

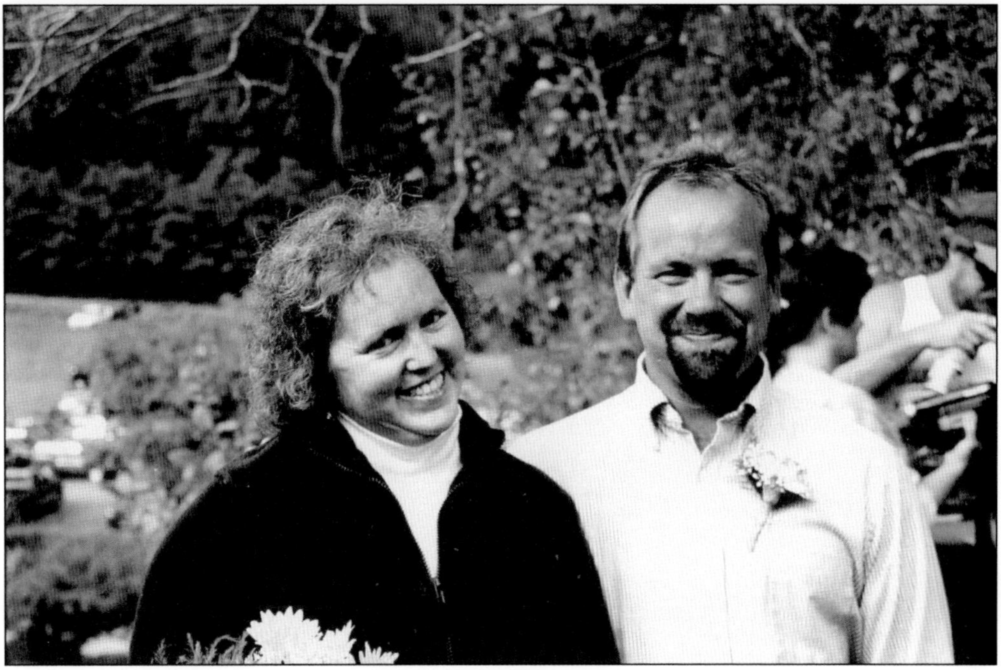

Saturday afternoon during the 2000 fair, Debra Mullen and Shane Craig were married by Euclid Farnham (a justice of the peace). Mullen wanted the ceremony to be as close to the cattle barns as possible because several generations of her family have been ox teamsters. (Courtesy of the Valley News; photograph by Amy Thompson.)

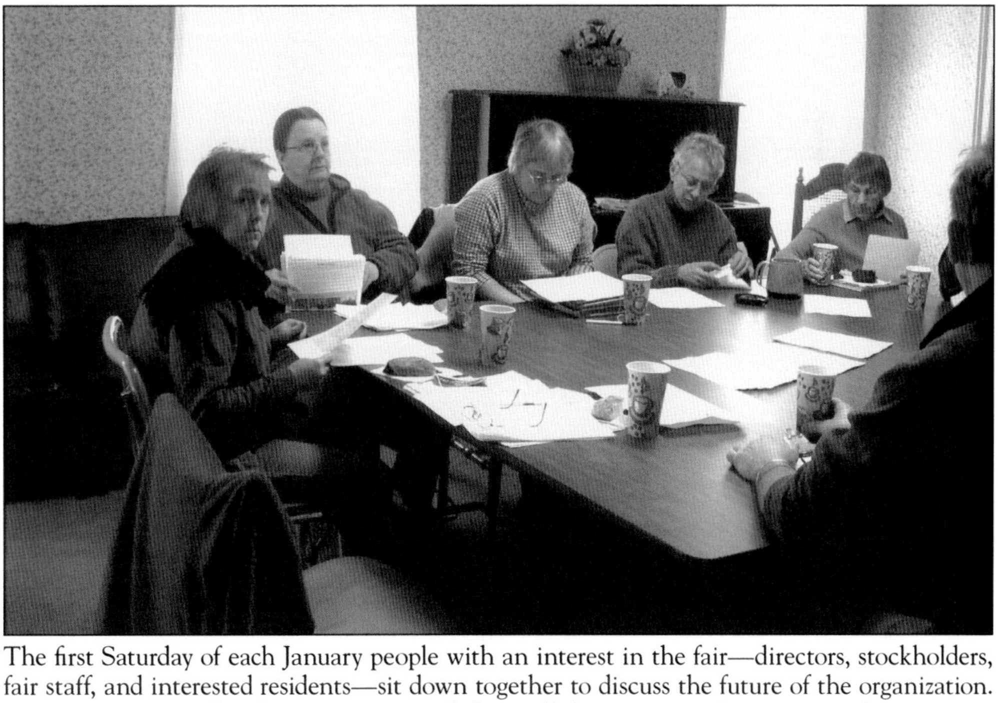

The first Saturday of each January people with an interest in the fair—directors, stockholders, fair staff, and interested residents—sit down together to discuss the future of the organization. Many innovative ideas have come out of these all-day meetings. Long-range planning is a necessity nowadays.

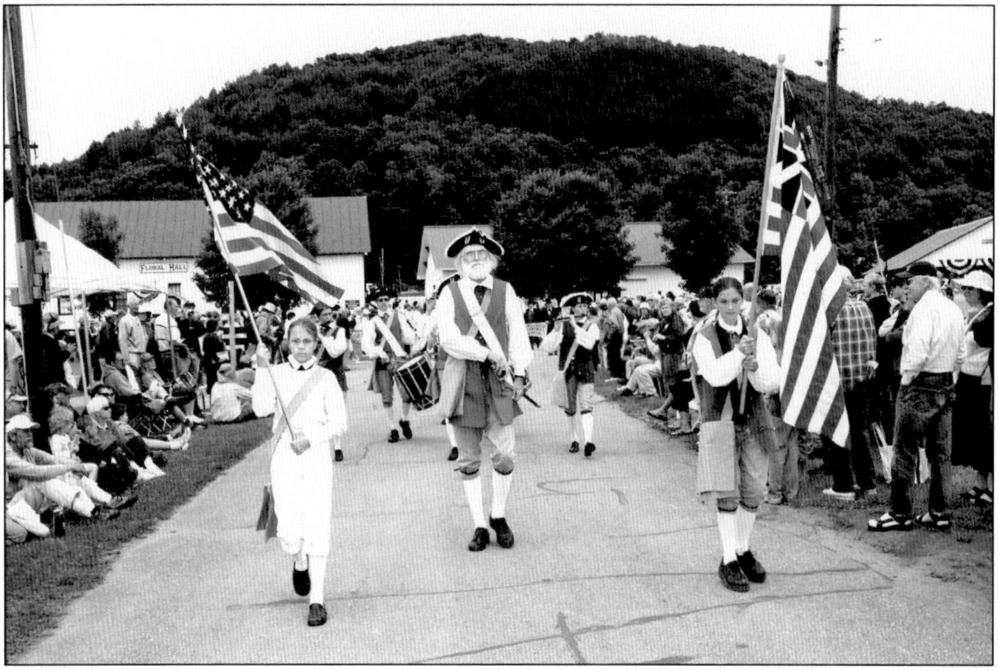

The Union Agricultural Society rents the grounds during the summer to various organizations. The following are some of the groups that utilize the property: Vermont History Expo (Vermont Historical Society), weddings, numerous horse shows and clinics, RV rallies, 4-H club gatherings, Animal Power weekend, trail rides, and martial arts training clinics. Above is the Vermont History Expo grand parade. (Courtesy of the Vermont Historical Society.)

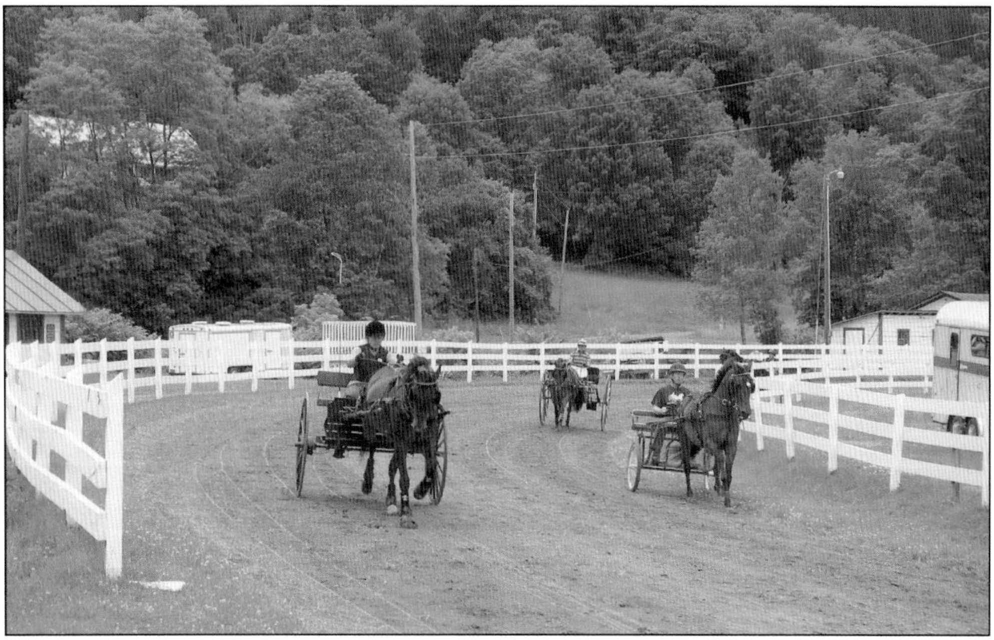

The state horse is the beautiful Vermont Morgan. Two Morgan shows are held yearly: the Morgan Heritage Days and the Lippitt Morgan Country Show. Icelandic horse associations also now are putting on summer meets. (Courtesy of Morgan Heritage Days.)

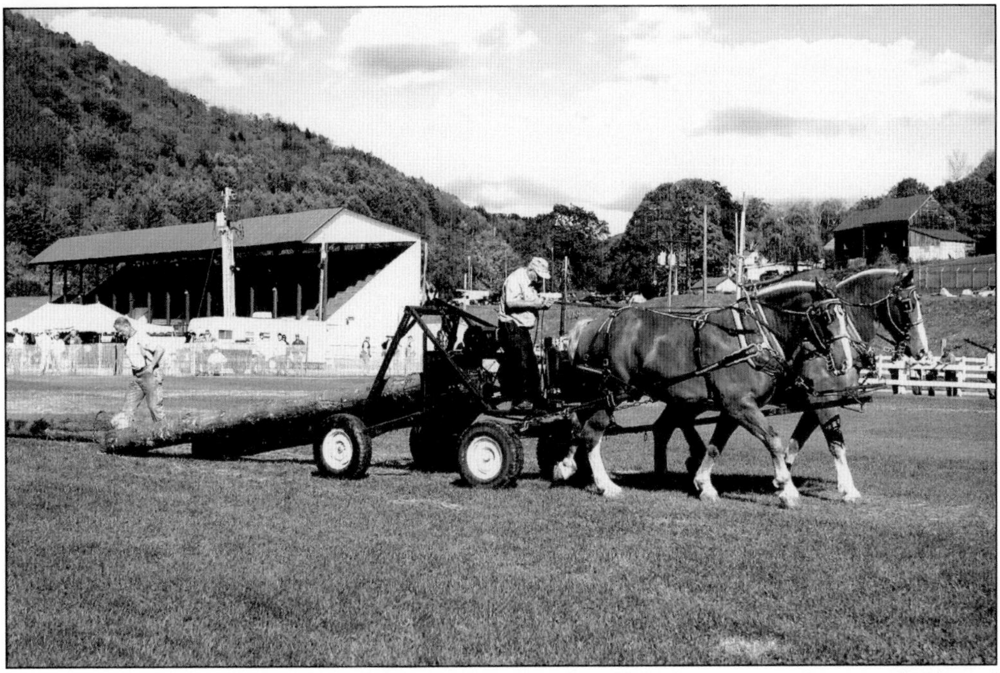

The Northeast Animal Power Field Days for a weekend in 2007 demonstrated new methods and equipment utilizing horses, mules, and oxen in forestry and farming. The first year was a great success. (Courtesy of Carl Russell, Northeast Animal Power Field Days.)

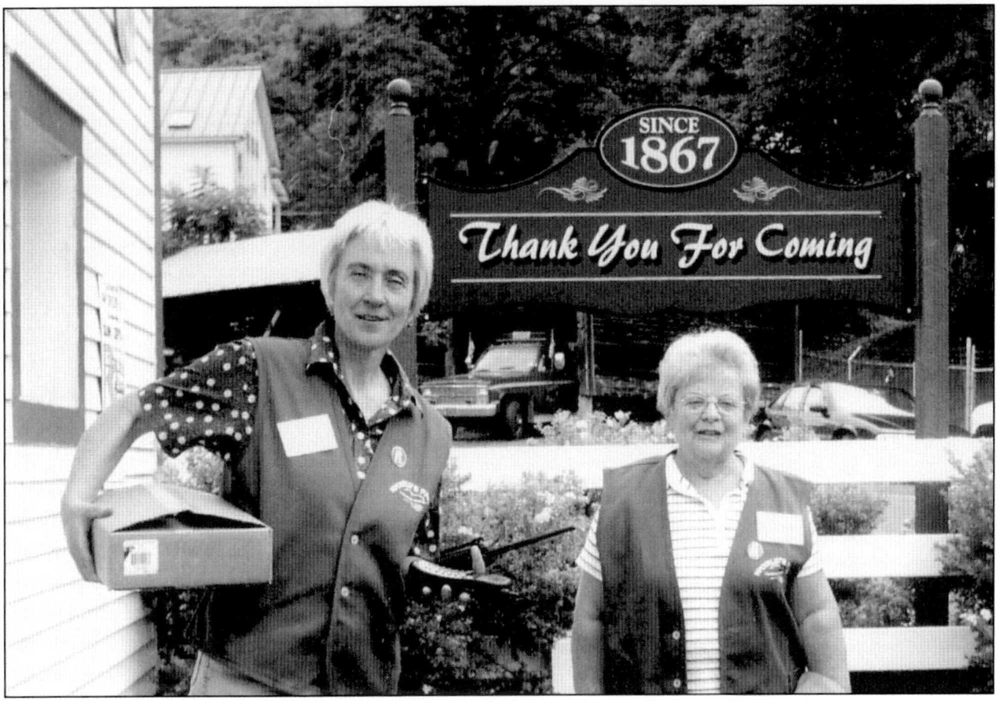

Administrative assistant Rebecca Béguin, left, proficiently handles multitudinous tasks: communications, graphic design, advertising, contracts, reports, and a daunting number of details. Joyce Miller, right, helps her run the Main Office smoothly, especially at fair time.

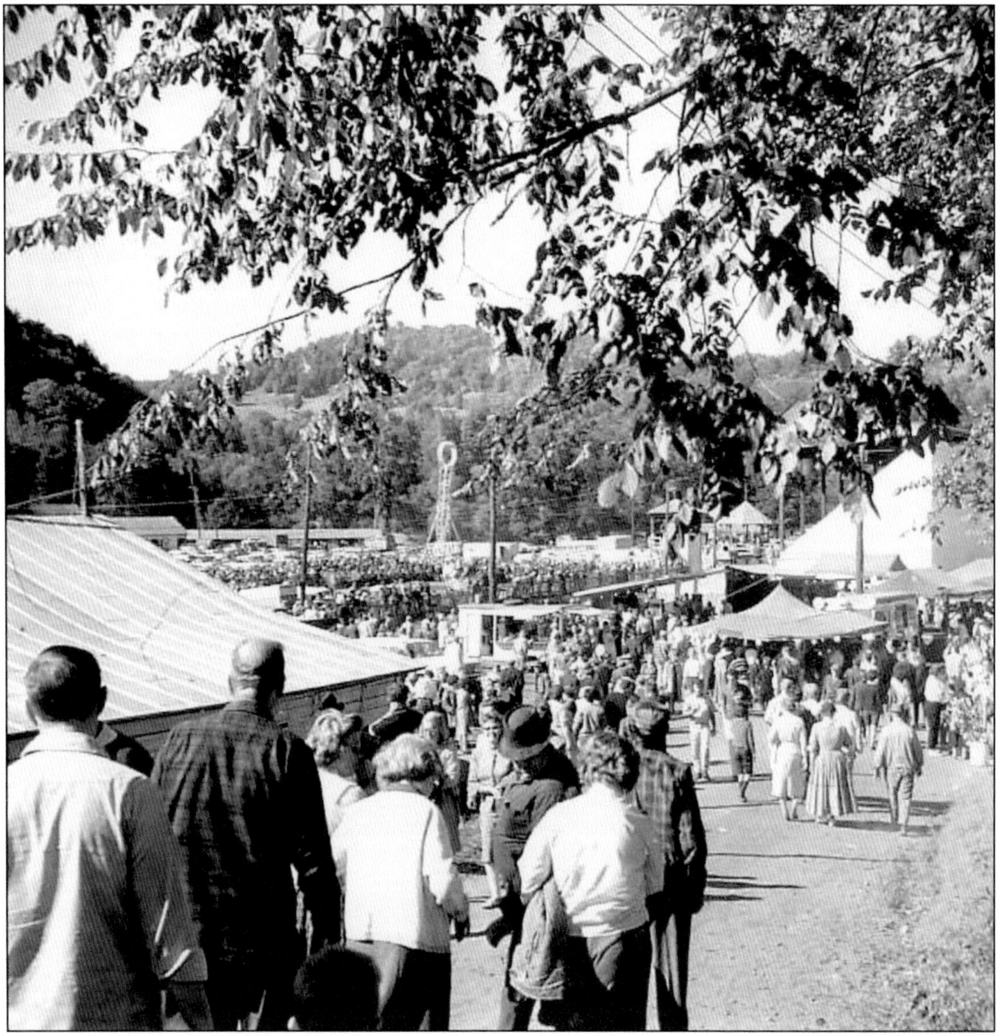

In the center of this 1950s scene is the rigging of a big wire acrobatic act. On the oval, thick crowds of people watch the outdoor pulling contest. The midway is jammed with people.

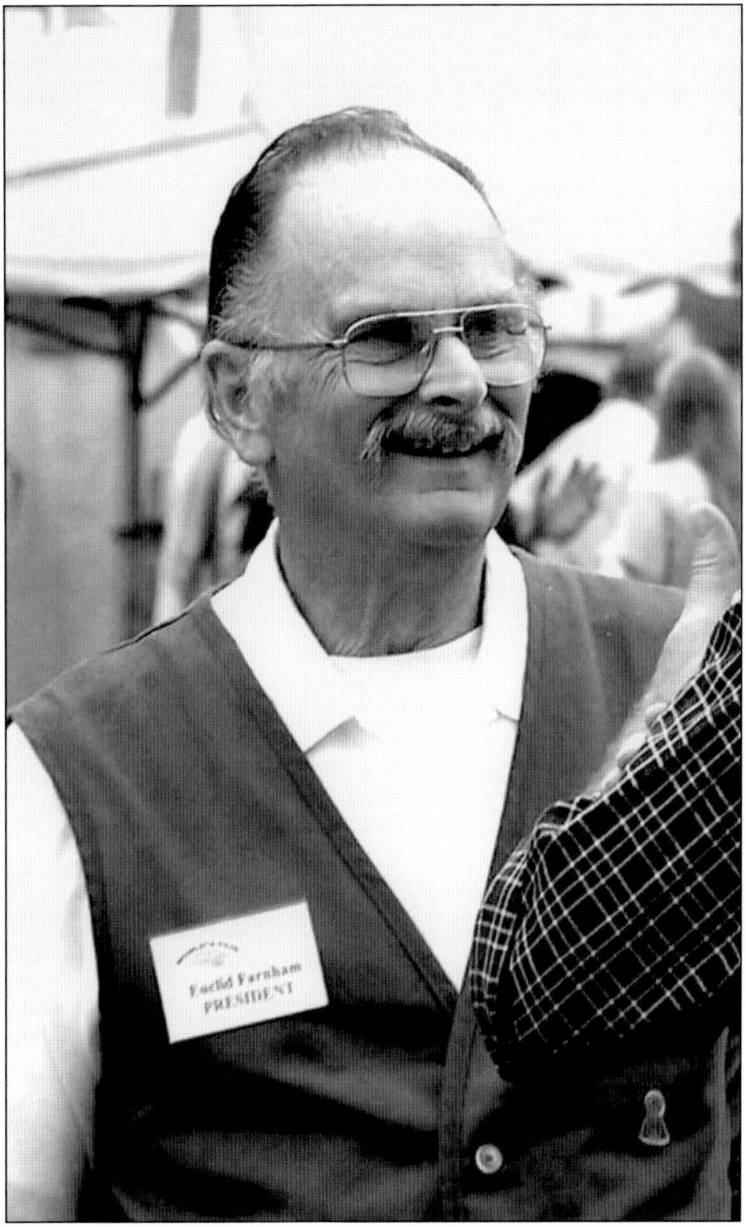

Euclid Farnham was born to William and Alice Farnham and raised on their dairy farm in Tunbridge. He first went to the fair with his mother just before he was born in 1933. Euclid well remembers his childhood delight that fair time brought: away from the hard work of farming, tail-gate picnics, a little pocket money for amusements, and above all, some free time. As a young man he took over the Jersey farm from his parents, retiring in 1995. Euclid also has devoted his life to his town and the study of history. He authored *Tunbridge Past*, a pictorial history of the town, and currently is working on another history of the area. He has greatly contributed to his town and fair in numerous ways: justice of the peace, town moderator for 27 years, church officer, treasurer of a number of organizations, photographer of Vermont and United States Civil War sites, fair director since 1969 and president since 1978, Vermont Fair Association president, and much more. (Courtesy of Marie Danforth.)

Consistent with our mission to preserve history on a local level, this book was printed in South Carolina on American-made paper and manufactured entirely in the United States. Products carrying the accredited Forest Stewardship Council (FSC) label are printed on 100 percent FSC-certified paper.

MADE IN THE USA